W0018077

The Global Politics of Human Embryonic
Stem Cell Science

Health, Technology and Society

Series Editors: **Andrew Webster**, University of York, UK and **Sally Wyatt**, Royal Netherlands Academy of Arts and Sciences, The Netherlands

---

Health, Technology and Society
Series Standing Order ISBN 978-1-4039-9131-7 hardback
(*outside North America only*)

You can receive future titles in this series as they are published by placing a standing order. Please contact your bookseller or, in case of difficulty, write to us at the address below with your name and address, the title of the series and the ISBN quoted above.

Customer Services Department, Macmillan Distribution Ltd, Houndmills, Basingstoke, Hampshire RG21 6XS, England

---

The Global Politics of Human Embryonic Stem Cell Science

Regenerative Medicine in Transition

by

Herbert Gottweis
University of Vienna, Austria

Brian Salter
King's College London, UK

and

Catherine Waldby
University of Sydney, Australia

© Herbert Gottweis, Brian Salter and Catherine Waldby 2009

All rights reserved. No reproduction, copy or transmission of this publication may be made without written permission.

No portion of this publication may be reproduced, copied or transmitted save with written permission or in accordance with the provisions of the Copyright, Designs and Patents Act 1988, or under the terms of any licence permitting limited copying issued by the Copyright Licensing Agency, Saffron House, 6-10 Kirby Street, London EC1N 8TS.

Any person who does any unauthorized act in relation to this publication may be liable to criminal prosecution and civil claims for damages.

The authors have asserted their rights to be identified as the authors of this work in accordance with the Copyright, Designs and Patents Act 1988.

First published 2009 by
PALGRAVE MACMILLAN

Palgrave Macmillan in the UK is an imprint of Macmillan Publishers Limited, registered in England, company number 785998, of Houndmills, Basingstoke, Hampshire RG21 6XS.

Palgrave Macmillan in the US is a division of St Martin's Press LLC, 175 Fifth Avenue, New York, NY 10010.

Palgrave Macmillan is the global academic imprint of the above companies and has companies and representatives throughout the world.

Palgrave® and Macmillan® are registered trademarks in the United States, the United Kingdom, Europe and other countries

ISBN-13: 978-0-230-00263-0   hardback
ISBN-10: 0-230-00263-3   hardback

This book is printed on paper suitable for recycling and made from fully managed and sustained forest sources. Logging, pulping and manufacturing processes are expected to conform to the environmental regulations of the country of origin.

A catalogue record for this book is available from the British Library.

Library of Congress Cataloging-in-Publication Data
Gottweis, Herbert, 1958–
   The global politics of human embryonic stem cell science :
   regenerative medicine in transition / Herbert Gottweis, Brian
   Salter, and Catherine Waldby.
       p. ; cm. — (Health, technology, and society)
   Includes bibliographical references and index.
   ISBN-13: 978-0-230-00263-0 (hardback)
   ISBN-10: 0-230-00263-3 (hardback)
   1. Embryonic stem cells—Research—Cross-cultural studies. 2. Human
   embryo—Research—Cross-cultural studies. 3. Regenerative
   medicine—Cross-cultural studies. 4. Cloning—Cross-cultural
   studies. I. Salter, Brian, 1946– II. Waldby, Cathy. III. Title.
   IV. Series.
   [DNLM: 1. Cloning, Organism—legislation & jurisprudence. 2.
   Embryonic Stem Cells—cytology. 3. Biomedical Research—ethics. 4.
   Public Policy. QU 328 G687g 2009]
   QH588.S83G68 2009
   174.2'8—dc22   2008037745

10   9   8   7   6   5   4   3   2   1
18   17   16   15   14   13   12   11   10   09

Printed and bound in Great Britain by
CPI Antony Rowe, Chippenham and Eastbourne

Contents

Tables

Acknowledgements

Herbert Gottweis is grateful for support of his research by the EU 6th Research Framework Programme (FP 6-SSH) project PAGANINI (Participatory Governance and Institutional Innovation), and he and Brian Salter acknowledge the support by the EU 7th Research Framework Programme (FP 7-SSH) project REMDiE (Regenerative Medicine in Europe: Emerging Needs and Challenges in Europe). Gottweis is also grateful for the research support by Ingrid Metzler, Byoungsoo-Kim and Robert Triendl. Herbert Gottweis, Brian Salter and Catherine Waldby also acknowledge the support of the UK Economic and Social Research Council Stem Cell Initiative Project Grant 'the Global Politics of Human Embryonic Stem Cell Science' (RES-340-25-0001). Catherine Waldby also acknowledges the generous support of the Sydney University International Research Fellowship Programme.

Series Editors' Preface

Medicine, health care, and the wider social meaning and management of health are undergoing major changes. In part this reflects developments in science and technology, which enable new forms of diagnosis, treatment and the delivery of health care. It also reflects changes in the locus of care and burden of responsibility for health. Today, genetics, informatics, imaging and integrative technologies, such as nanotechnology, are redefining our understanding of the body, health and disease; at the same time, health is no longer simply the domain of conventional medicine, nor the clinic.

More broadly, the social management of health itself is losing its anchorage in collective social relations and shared knowledge and practice, whether at the level of the local community or through state-funded socialized medicine. This individualization of health is both culturally driven and state sponsored, as the promotion of 'self-care' demonstrates. The very technologies that redefine health are also the means through which this individualization can occur – through 'e-health', diagnostic tests and the commodification of restorative tissue, such as stem cells and cloned embryos.

This Series explores these processes *within* and *beyond* the conventional domain of 'the clinic', and asks whether they amount to a qualitative shift in the social ordering and value of medicine and health. Locating technical developments in wider socio-economic and political processes, each text discusses and critiques recent developments within health technologies in specific areas, drawing on a range of analyses provided by the social sciences. Some will have a more theoretical, others a more applied, focus, interrogating and contributing towards a health policy. All will draw on recent research conducted by the author(s).

The Health, Technology and Society Series also looks towards the medium term in anticipating the likely configurations of health in advanced industrial societies and does so comparatively, through exploring the globalization and the internationalization of health, health inequalities and their expression through existing and new social divisions.

This book makes a valuable contribution to the Series by offering a detailed examination of the global dynamics shaping the contemporary biosciences and the relationship between the 'knowledge economy' and the regulatory state that shapes their development. The book fills an important gap by providing a thorough analysis of the global regulation of regenerative medicine both with respect to embryonic stem cells and more mature tissue technologies. The core economic, sociological, political and especially (globalized) regulatory aspects of regenerative medicine are described in a very clear and coherent way, building towards a conclusion that flags up the future challenges for state and society at both local and global levels. Throughout the text there is an ongoing anchorage within the wider globalization literature, so the book will be of value to researchers and students working in both the specific area of the regulatory dynamics of regenerative medicine and the wider field of the globalized knowledge economy.

Andrew Webster and Sally Wyatt

Introduction: Stem Cell Research and Global Biopolitics

On 11 August 2004 Miodrag Stojkovic and Alison Murdoch of the University of Newcastle upon Tyne received the United Kingdom's first licence to create human embryonic stem cells (hESCs) using cell nuclear replacement. Stojkovic and Murdoch's application was based on research funded by the UK Department of Health, the Department of Trade and Industry and a regional development agency. When the news from Newcastle circulated around the world, reactions varied. A spokesman for the German Ministry of Research and Education stated that a scientist based in Germany would be committing a crime, simply by the act of advising a London colleague by telephone on the creation of embryonic stem cells (ESCs). In contrast, in California, on 2 November 2004, a statewide ballot authorized the government to spend up to $3 billion over 10 years on stem cell research – a plan intended as a direct assault on President George W. Bush's strict limits on ESC research. At the Third Annual Meeting of the International Society for Stem Cell Research in San Francisco in June of the following year, Robert N. Klein, the author of the California Stem Cell Research and Cures Initiative and the father of a child suffering from juvenile diabetes, delivered the keynote address to an auditorium of scientists, representatives from industry and policymakers who celebrated Klein as a political visionary.

Rather more dramatically, in October 2005, South Korean president Roh Moo-hyun officially launched the World Stem Cell Hub – the brainchild of ESC pioneer Hwang Woo-Suk, who had written a paper published in *Science* that spring describing the creation of 11 patient-specific stem cell lines (Hwang et al., 2005). The new organization's clearly stated goal was South Korean leadership in global stem cell research and its medical applications. Negotiations were under way to establish regional subsidiaries of the facility in a number of countries, including the United States, the United Kingdom and Germany. At the same time, Hwang was busy submitting patent applications connected to his work in Europe and in the United States. Barely 4 months later, in January 2006, South Korea's ambitious stem cell research programme was in tatters following confirmation that the 11 claimed cell lines were crude fakes and that Hwang had committed gross ethical violations in securing egg donations. Hwang was disgraced and his paper was retracted by the editor of *Science* (Kennedy, 2006). South Korea's science minister and the chairman of the country's newly created Bioethics Committee were forced to resign after being implicated in the Hwang scandal. Hwang's disgrace sent shockwaves around the world. Commentators questioned the validity of scientific peer review and urged reform, accusing journal editors of sacrificing judicious assessment of manuscripts for the next big story. Scientists who had collaborated with Hwang nervously checked their records.

At first sight, these four scenes from the world of stem cell research seem to be unconnected. But there are, in fact, important relationships between them. Stojkovic, who in May 2005 announced that he had successfully produced an early-stage embryo cloned from a human cell using nuclear transfer, had moved earlier from the University of Munich, Germany, to the United Kingdom to be able to do work in hESC research. With the United Kingdom striving for world leadership in hESC research, proponents of California's Stem Cell Initiative had warned that the United States would soon lag behind in stem cell research, reducing its appeal to scientists and potential industrial investors. And Hwang, for several years the 'King of Cloning' in the field of stem cell science, had not only shown his global ambitions through the creation of the World Stem Cell Hub but also tried to secure his apparent leadership through wide-ranging, worldwide patent applications. If anyone had doubted it before, stem cell science was now clearly shown to be a global enterprise, interconnected and interdependent at many and various levels.

In this book, we investigate these global relationships between the commercial and scientific trajectory of stem cell research; the contentious ethical status of embryos and stem cells; the cultural tensions they generate; the global biopolitical relations they have helped to create; and the complexities of regulation they present for national, regional and international governance bodies. In doing so, we hope to contribute to current debates and understandings of the social implications of stem cell research in particular and contemporary regenerative medicine policy and innovation more generally. By focusing on interactions between local, national, regional and global politics and governance, we will frame regulatory regimes within a network of global dynamics, rather than regard them as the products of discrete national histories and debates. We also hope to contribute to a better understanding of the interrelations between the multifarious social, economic, cultural and technological forces described by the term *globalization* in the arena of regenerative medicine, and more generally, the life sciences. Our book has a clear focus on stem cell research, but we also tell a story about regenerative medicine, a newly emerging field of medical research and application. Although stem cell research, particularly hESC research, has a number of highly specific characteristics, we believe that many of the insights we present here are shared by other areas in regenerative medicine. Regenerative medicine, the combined effort of medical and biotechnological approaches to develop smart scaffolds and molecules to direct tissue regeneration in patients and eventually to develop 'custom-made' tissue and organs, raises a highly specific set of shared political, ethical, regulatory and legal challenges, and a new global politics seems to be emerging in this field.

In particular, we focus on the tight interrelationships among the shaping of new knowledge regimes in regenerative medicine, their financial backing by a variety of state and non-state actors, the often contentious status of the new knowledge regimes, attempts to stabilize and interconnect the various components of the new knowledge regimes through standardization and efforts to establish new appropriations regimes, typically through patenting. The combination of these different developments gives rise to new forms and mixes of private/public governance in regenerative medicine on a global level.

The therapeutic promise of ESCs

Since the development of safe blood transfusion in the 1930s, the redistribution of human tissues from a donor to a recipient has been a lifesaving medical technology. Today almost all the tissues in the human body – solid organs, skin, bone, heart valves, corneas – can be transplanted in some form, and complex donation, storage and triage systems have been developed to circulate these vital materials. HESCs are the most recent innovation in transplantable tissues, so recent that their clinical application is yet to be demonstrated. They are also highly complex and controversial entities, rarely out of newspaper headlines since their development in 1998.[1] As their name implies, hESCs are a form of tissue derived from human embryos, in their early stages of development. Researchers believe that hESCs have particular qualities that may make them highly desirable therapeutic agents. In theory, a stem cell line is a self-renewing tissue resource, and if stem cell lines can be moved from experimental to therapeutic status, they may be able to compensate for acute shortages in many areas of the human tissue economy (Waldby and Mitchell, 2006; Waldby, 2002) as the demand for donated organs, blood and cellular material steadily outstrips supply. Stem cell lines may also be able to assist in the treatment of currently untreatable degenerative conditions such as Alzheimer's disease and spinal cord injuries. For states in the developed world struggling to manage ageing populations with increased rates of chronic disease and disability and ever-growing expenditures on health care, stem cell technology is an especially attractive development. Commercial biotechnology firms such as Geron and ES Cell International are also engaged in active research programmes, with the expectation of developing stem cell technologies and therapies for what might one day be a global health market.

At the same time, hESC technologies have proved to be controversial throughout the world. Although blood, organs and cellular material are historically donated by adult citizens who formally consent to give part of their bodies to another, embryonic material has a somewhat different status. Human blood or organs are ethically significant because they *refer to or derive from* a person whereas many people consider that embryos *constitute* persons.

Much of the debate on cloning and stem cell research since 1997 seems to be not only about human life and its beginnings but also about acceptable strategies for controlling the creation and development of human life. In fact, the various strategies to obtain ESCs and their application in medical research and therapy have turned immediately into a site of political controversy in Europe, the United States, Australia and other countries. Many issues and topics are raised in these debates, including the rights of embryos; the question of the potential risks for patients involved in these novel, experimental medical strategies; and the implications for women because of the possible increase in demand for 'spare embryos', oocytes and other reproductive tissues. With the emergence of hESC and cloning research, old lines of argument from former controversies about fetal research and abortion were reactivated. At the same time, new interpretations, facts and styles of controversy affected the construction of the meaning of hESC research. Even today when taking a cursory look at the stem cell/cloning debates in most industrialized countries, one might come to the conclusion that society and the state are faced with critical choices about the future of humankind.

Because of the tensions involved within this highly emotional and conflict-filled debate among industry, state interests, consumer advocates and ethical opponents, medical research involving human embryos takes place in a heavily regulated and sometimes bitterly contested environment in many countries. Policies range from the complete prohibition of hESC research (as in Ireland, Austria, Lithuania, Poland and Slovakia) through an array of regulatory configurations that allow certain kinds of research to policies that permit the creation of embryos for research and for therapeutic cloning (as in Belgium, India, Israel, Sweden, China, Singapore, South Korea and the United Kingdom). Each set of regulations is the product of protracted negotiation, argument, stand-off and compromise. Each regulatory configuration reveals complex, often conflicting understandings of the status and meaning of prenatal life and the legitimacy of technical interventions into fundamental aspects of human reproduction and thus human origin. The regulatory domain in stem cell science has become a theatre where the biopolitical narratives and networks of science, industry, the state and civil society play out around philosophical questions about the status of the human, the claims of the aged and disabled, the power of biotechnology, and the biological future of society and the species. Although questions of the status of embryos have been important in this context, other topics such as the transformation of cells into therapeutic agents through genetic modification and the potential move from therapy to 'enhancement' are raised by the practices and promises of regenerative medicine.

Regenerative medicine has also become a *global* theatre. Although the regulation of biotechnology is carried out primarily by the governments of nation states, the biopolitical forces and technological vectors that shape stem cell science exceed the boundaries of the nation state and involve regional, transnational and global alliances, material flows, corporate structures and civil associations. Patient-advocacy groups are able to take advantage of global communications technology to organize their lobbying activities, as are Christian coalition members and right-to-life advocates. Stem cell research, like much other biotechnology research, takes place on a global scale, in both the public and private sectors. Companies, laboratories and scientists are highly mobile actors, able to relocate in response to liberal regulatory environments and offers of public funding. For example, Alan Coleman, former research head of PPL Therapeutics in Edinburgh, moved to Singapore in 2002 to take advantage of the generous public research funding for stem cell research through the Biomedical Sciences Initiative and the creation of the Biopolis life sciences research complex. The embryos used to create stem cell lines are traded around the world, exported by countries such as Israel and India, with large in vitro fertilization (IVF) programmes and few bioethical restrictions, to countries such as Germany and Singapore, with high levels of scientific expertise but limited sources of 'spare' embryos (Bharadwaj and Glasner, 2004; Prainsack, 2004). Stem cells themselves are eminently global entities: the same stem cell line can be multiplied, subdivided, dispersed and utilized in laboratories throughout the world, and each cell line has an indefinite, and theoretically immortal, lifespan.

Stem cell technologies thus form what Arjun Appadurai terms *global technoscapes*: 'the global configuration ... of technology, both high and low, both mechanical and informational, [which] moves at high speeds across various kinds of previously impervious boundaries' (Appadurai, 1996, p. 34). Stem cell technology forms part of the global life sciences economy, linking investment capital, scientific expertise and biological material in profitable, transnational configurations. The life sciences are also exemplary of the mobile, hybrid entities whose global circulation is helping to reshape social relations. As Urry (2000) states, the social processes we associate with the term *globalization* – high levels of global interdependence, mass mobility of people and objects, instantaneous communications and financial transactions, the weakening of the nation state and the complication of jurisdictional powers – are mediated by 'inhuman objects'. '[Social relations] are made and remade through machines, technologies, objects, texts, images ... and so on. Human powers increasingly derive from the complex interconnections of humans with material objects' (p. 14). In this sense, the global circulation of stem cell technologies and other biotechnical entities have helped to create new forms of alliance and opposition, new kinds of bodily exchange and indebtedness (Waldby, 2002; Waldby and Mitchell, 2006), new experiences of belonging and identity (Rose and Novas, 2005) that transcend, or are detached from, the regulatory space of the nation state and national citizenship.

This global scale and the contentious nature of stem cell research mean that new transnational and public/private forms of regulation are beginning to emerge in the stem cell arena. The various governance and deliberative organs of the European Union (EU) have acted as regional forums for debates and policymaking around stem cell technology, the status of prenatal life, the rights and interests of patients and the imperatives of scientific development. The EU's European Group on Ethics has proved to be a particularly important venue for deliberation over the biopolitics of stem cell technology, whereas the EU Framework Programme, which provides public research funding, has been the site of fierce lobbying, stalemate and compromise over stem cell research funding. States with an interest in stem cell research have begun to use the United Nations (UN) to battle for control of the international agenda, especially over the most contentious area of stem cell research, the issue of embryo cloning. International bioethical conventions, such as the Council of Europe's Convention on Human Rights and Biomedicine (1997) and UNESCO's Universal Declaration on the Human Genome and Human Rights (1997) are becoming more prominent in shaping national debates and regulatory policy. Subnational sites also play an emerging role in the global stem cell economy and debate. In the United States, where electoral politics and the conservative Bush administration have placed strong restrictions on publicly funded stem cell research at the federal level, several state administrations have provided their own legislation and significant public funding to develop stem cell research.

Stem cells, life sciences and the new knowledge economies

Stem cell research is part of the explosion of life sciences innovation taking place in the 'knowledge economies' of the developed nations. These new life sciences knowledge economies are the result of a recent reorganization of the global economy and developments that have occurred in basic biological research since the 1950s. As we discuss in detail in Chapter 1, the contemporary era of economic globalization is widely attributed to the dramatic decline in profitability of the nationally based, Fordist modes of mass manufacture that underpinned the post-war economic boom. Corporations seeking to retool from unprofitable mass production to post-Fordist 'flexible accumulation' have turned, since the late 1970s, to new scientific knowledge as a possible new source of value. This is particularly true of what Etzkowitz and Webster (1995) term the *enabling technologies* – information, communications and bioscience technologies – 'that underpin a wide range of industrial sectors without being unique to any one ... [and] upon which a growing number of sectors must rely to remain competitive' (p. 495). Corporations seeking knowledge with commercial applications have typically formed research partnerships with university departments seeking supplementation for dwindling public funds. Such relationships, fostered at a policy level by the Bayh-Dole Acts in the United States, facilitate corporate flexibility through outsourcing specialist knowledge production while they represent a source of funding and prestige for universities forced off public sources of funding by contracting state budgets (Etzkowitz and Webster, 1995). The extension of intellectual property (IP) rights to scientifically engineered living organisms has also been a crucial factor in fostering commercial investment in life sciences research. These developments have produced an explosion of innovation in the life sciences industries (pharmacology, agribusiness, medical biotechnology), organized around new sets of genetic and cellular techniques to manipulate life. To use Waldby's term, biological life is being transformed into *biovalue*.

> Biovalue refers to the yield of vitality produced by the biotechnical reformulation of living processes. Biotechnology tries to gain traction in living processes, to induce them to increase or change their productivity along specified lines, intensify their self-reproducing and self-maintaining capacities. This intensification or leveraging of living process typically takes place not at the level of the body as a macro-anatomical system but at the level of the cellular or molecular fragment, the mRNA, the bacterium, the oöcyte, the stem cell. Moreover it takes place not *in vivo* but *in vitro*, a vitality engineered in the laboratory ... Here a repertoire of biotechnical procedures can be developed that induce the fragment to expand, to accelerate or slow down, to unfurl or recapacitate, to produce new substances or develop along new pathways, to recombine with other fragments and swap properties. In short biotechnology finds insertion points between living and nonliving systems where new and contingent forms of vitality can be created, capitalizing on life.
>
> (Waldby, 2002, p. 310)

Today biotechnology firms such as Geron, Aastrom Biosciences and ES Cell International and agribusiness firms such as Monsanto reengineer living processes as their core business, and they constitute a significant element in the economies of the developed world.

Stem cell research has emerged from a web of such biovaluable innovation and from the failure of genetically based technologies to produce therapeutic outcomes with clinical applications. From the late 1950s until the mid-1990s, the manipulation of DNA has been the dominant focus of biomedical attention. Genetic engineering, genetic testing, gene therapy and the human genome project seemed to delineate a new future for biomedicine in which knowledge and mastery of DNA were the keys to solving many of the important problems in medical research (Nelkin and Lindee, 1995). During the 1990s, however, it became clear that the medical claims made for genetics would not be realized. The human genome project greatly helps the growth of the genetic testing industry but offers little in the way of therapeutics. New lines of biological and medical reasoning, often termed *post-genomic*, have gained ground. In particular, medical research concerned with the interaction between cell proteins and genes (proteomics) (Keller, 2000) and with developmental biology, embryology and reproductive medicine has come to the therapeutic forefront. Although these biomedical disciplines have pursued rather different research agendas throughout most of the twentieth century, more recently they have converged (Brown and Webster, 2004) in relation to stem cell technology in particular, and in the field of regenerative medicine in general.

Stem cells: Biology and history

Stem cells are cells that generate or regenerate tissue. They are undifferentiated cells that can renew themselves and also give rise to one or more specialized cell types with specific functions in the body. Stem cells exist during early organism development, but they also occur in adult tissues. They are abundant in the bone marrow, the developing brain and in two areas of the adult central nervous system: the hippocampus and the olfactory bulb. Mammals appear to have some 20 major types of somatic stem cells that, for example, can generate muscle, blood, intestine, liver and heart (McKay, 2000). In human beings, the skin, gut, blood and uterine linings are regularly replaced by stem cell activity, and stem cells play a crucial part in wound healing. The most celebrated type of stem cell is 'pluripotent', meaning that it has the capacity to develop into almost all of the body's tissue types. Recent research suggests that it may be possible to produce large numbers of pluripotent stem cells that differentiate on demand, providing an unlimited supply of transplantable tissue. Medical researchers think stem cells may be very useful in treating currently intransigent medical conditions – Parkinson's disease, Alzheimer's disease, stroke, spinal cord injuries, arthritis – through the introduction of tissue into damaged sites. Stem cells might provide alternative therapies for common conditions such as diabetes, promoting the growth of insulin-producing tissue to replace pharmaceutical regimes. The cells may act as substitutes for organ donation, repairing an existing heart or kidney rather than replacing it. Moreover, it may be possible to produce stem cell lines that are genetically and immunologically compatible with particular hosts, avoiding the problem of tissue typing found in organ transplants (McLaren, 2000; Vogel, 2004a).

Human embryos are a key source of pluripotent stem cell lines. Other stem cell sources such as induced pluripotent stem (iPS) cells gained from human skin have recently received much attention. The groups of Shinya Yamanaka at the University of Kyoto and James Thomson in Wisconsin had used genes to reprogramme human cells so that they had all the characteristics of hESCs – but without being derived from human embryos (Takahashi et al., 2007; Yu et al., 2007). This development seems to offer a novel approach towards regenerative medicine, even with the potential of creating oocytes and sperms with pluripotent stem cells in the middle-range future. However, whether iPS cells can indeed fully substitute for ESCs still remains to be seen (Cyranoski, 2007).

ESCs were first isolated from the inner cell masses of mouse blasto-cysts in the early 1980s. Around the same time, R. G. Edwards (one of the pioneers of in vitro fertilization (IVF) technology) and colleagues grew the first human blastocyst in vitro 5 days after insemination. Both Edwards's work and the work on mouse embryos were closely related to the growth of reproductive medicine and the development of IVF since 1962 (Edwards, 2001).

Breakthroughs in hESC research remained closely tied to the expan-sion of IVF technologies. In 1998 a team of researchers under John Gearhart at Johns Hopkins University managed the first successful isolation and culturing of human embryonic germ cells, derived from aborted foetuses (Gearhart, 1998). At about the same time, a University of Wisconsin research team led by James Thomson derived ESCs from spare embryos created in IVF clinics for reproduction and donated by couples who no longer needed them. The research-ers grew each embryo for about 5 days until it had developed into a blastocyst, and they isolated the inner cell mass and established five immortalized cell lines[2] (Thomson et al., 1998). Thomson's team cultured the cell lines for 5 months without differentiation and then induced the lines to differentiate into the main groups of embryonic tissue layers. The cell lines could also be frozen without apparent damage to their replicative capacities. The Thomson group's research was widely interpreted as a crucial step in the development of new strategies to grow human tissues and organs (Cohen, 1998). The new research was nominated by *Science* as among the top scientific advances of 1999 (Vogel, 1999).

These breakthroughs in stem cell science were rendered all the more exciting and therapeutically promising by their intersection with the advances in mammalian cloning technology made famous by the birth of Dolly, the sheep, in Britain. In 1997 Ian Wilmut and other scientists from the Roslin Institute and PPL Therapeutics announced the first successful mammalian cloning, using a technique known as somatic cell nuclear transfer (SCNT). SCNT involves the creating of an embryo not by the usual process of in vivo conception, fusion of egg and sperm, but through the in vitro insertion of the nucleus of a cell from an adult's tissues into an oocyte, an unfertilised egg. The oocyte has in turn been enucleated – had its own nucleus removed to make way for the introduced nucleus. This creates an embryo with the genome of the adult from whom the nucleus was taken. Prior to the cloning of Dolly, it was assumed that the nuclei of adult cells had lost their pluripotency, that is, once programmed to produce a particular kind of cell, they lost their ability to produce different kinds of cells (Keller, 2000). Cloning based on SCNT demonstrated that adult cell nuclei could, in fact, be induced to revert to or reactivate their embryonic potential. In Dolly's case, in a practice termed *reproductive cloning*, this embryo was introduced into the uterus of a surrogate mother, who carried it to term, producing a genetic copy of the gene donor. Reproductive cloning has since been used to produce cattle, sheep, mice, goats and pigs.

Mammalian cloning expands the possible therapeutic repertoire of stem cell technology because it may be possible to use some aspects of this technique to develop ESC lines with the genetic material of an adult donor. Such a practice, known as therapeutic cloning, would produce transplantable tissues genetically compatible with the donor and so avoid the problem of immunological rejection that plagues all forms of tissue transplant. The embryonic tissue would be compatible with those of the donor and could be used to repair organs or regenerate tissues. Such tissues would carry no risk of rejection, and the person would not need to take immuno-suppressive drugs.

Stem cells, biomedicalization and health markets

The rise of stem cell research and cloning technologies resonated with a number of broader developments. HESC research is not only a scientific activity but also a practice of social and political ordering. Human tissues created with ESC technology are potentially lifesaving and reflect images of the acceptable boundaries between humans and humans, understandings of the rights of embryos and visions of what constitutes acceptable risks for citizens and for the society at large. As Schaeffer and Shapin (1984) have shown so convincingly, solutions to problems of knowledge are often solutions to problems of social and political order. In the medical field, one important expression of this phenomenon of multiple ordering has been subsumed under the term *medicalization*. Initially framed by Irving Zola (1972), the term implies the extension of medical jurisdiction, authority and practices into increasingly broad areas of people's lives. By conceptually redefining particular phenomena such as alcoholism and drug abuse in medical terms, medicine became a new agent of social control for managing deviance.

Today medicalization recognized as a coproduction of science, technology and new social forms has become known as *biomedicalization* (Clarke et al., 2003). Biomedicalization, located at the transition from the problems of modernity to the problems of postmodernity, is characterized by a shifting of the sites of control from external nature (i.e., the world around us) to the harnessing and transformative customizing of internal nature (i.e., biological and physiological processes of human and nonhuman life forms), often transforming life 'itself'. At the core of biomedicalization are the enhanced capacities of an increasingly technoscientifically constituted biomedicine to effect the transformation of bodies and lives. Such transformations range from giving birth a decade or more after menopause to the capacity of genetically redesigning life. At the same time, biomedicalization refers to a deep-seated institutional-organizational transformation of medicine, implying such diverse processes as the production of new social forms by the usage of computer and information technologies and the corporatization (e.g., HMOs (Health Maintenance Organizations)) and commodification of research (e.g., Celera Genomics), products and services in the medical system (Clarke et al., 2002). We can view stem cell and cloning research as being indicative of these processes of biomedicalization. Previously, the central goal of medicine was to arrest abnormality and to reestablish the vital norm. With the new cellular technologies and genetic subcellular strategies, the normativities themselves appear open to transformation. The line between correcting deviant genes that cause diseases and modifying genes to enhance athletic performance is a thin one. In a similar vein, stem cell approaches might help cure Parkinson's patients, but, in principle, they could also be used to improve brain functions. Life now seems to be open to shaping and reshaping, and the distinctions between treatment and enhancement, between the natural and the prosthetic, begin to blur (Rose, 2001, p. 15). The cellular therapies currently under development, such as therapeutic cloning or tissue engineering, seem to promise a vast expansion of the scope and depth of intervention in life processes. As the normativity of life begins to lose its contours, answers to the questions about who we are and what we can hope for have become ever more challenging. Already, ethicists ponder the implications of stem cell therapy in the human brain for personal identity.

The rising new strategies to 'rebuild life' must also be seen in the discursive context of current shifts in health policy. Since the last quarter of the twentieth century, challenges such as demographic development, access to and quality of care, cost crises, and technoscientific innovations like modern biomedicine share at least three phenomena that have become driving forces for policy reform in the health-care sector: (1) globalization, (2) the crisis of the welfare state and (3) the neoliberal narrative in health policy. They represent escalating pressure for standardization, increasing significance of free-market solutions and a managerial paradigm in most areas of social interaction, and an acceleration of both technological and scientific innovation and its application in medicine. Since the 1970s, national health-care systems have increasingly applied similar tools and developed institutional designs to cope with similar policy problems and challenges. These changes led to a redefinition of hitherto-stable policy patterns and fostered a global techno-managerial paradigm in health policy that stresses the importance of increased scientific input and efficacy in both the provision and organization of medicine. The retrenchment of the welfare state because of the global hegemony of a neoliberal narrative failed to reduce regulation and instead led to micromanagement and microregulation of patients' and physicians' behaviour by means of the private market. Health care is increasingly perceived and treated as similar to any commodity market (Scheil-Adlung, 1998). The neoliberal health-care discourse moved to a governmental style, in which the social takes the form of markets. This reassertion of the role of the market, in particular, serves as a mechanism for constructing identities as targets for managed-care action. The sick are regarded as consumers and physicians as entrepreneurs in the health-care market. The contrivance of markets became the technical means for the reformulation of all types or provision in the health-care market (Dean, 1999, pp. 171–2).

Today, health is increasingly discussed in terms of self-control and framed within the language of an ethics of health, which requires the disciplining of the individual conduct of life (Crawford, 1984, pp. 72–6) throughout the West. This managing of the self (Foucault, 1983 [1976], 1994 [1975]) is also reflected in a multitude of technical and organizational novelties within health care, where managed care is the most important and paradigmatic example. In the past, health policies were conducted based on the collection and tabulation of numerical information about populations and provided the rationale for hygienic strategies. Likewise, strategies to minimize environmental and workplace risks and improve the maintenance of the body were central elements in public health strategies. Although these strategies continue to be important, the focus has begun to shift from the group to the individual level. The ideal of the omnipresent state that would shape, coordinate and direct the affairs in all sectors of society has lost its grip on the public imagination. Accordingly, in the health field, focus has shifted from 'society as a whole' to 'risky individuals', individual susceptibility (to genetic disease, for example), and accordingly, to 'risk groups' (Rose, 2001). The proactive management of the human body has become a core element of collective and individual strategies of health maintenance.

Regenerative medicine and ageing populations

Closely related to the management of collective health is the question of the regeneration and reproduction of ageing populations with decreasing fertility rates. By 2020, approximately 20 per cent of the population in the G8 nations will be over age 65 (Neilson, 2003), and they will live longer than previous generations. As Neilson points out, this shift in the demographic profile of the industrial democracies presents a set of intractable policy problems to governments.

> With portentous consequences for the ratio of working-age taxpayers to nonworking retirees, these changes in age profile threaten the economic viability of the world's wealthiest and most powerful nation-states, tearing at the fabric of their once liberal notions of citizenship, constitutionalism, and social contracturalism ... population aging places a glacier-like pressure on the nation-state, slowly but surely eroding its centralized apparatuses for managing the production and reproduction of life.
>
> (2003, p. 163)

Ageing populations place large burdens on welfare and pension provisions at a time when governments in developed nations are threatened by electoral opprobrium if they increase tax rates. One answer to this problem would be to foster mass immigration from the more populous countries of the south to the depopulated countries of the north. However, many developed-nation governments have pursued anti-immigration policies for electoral gain and so closed off this option for population renewal.

States in the wealthy nations are therefore primarily committed to forms of internal renewal, and hESC technology offers two potential avenues for the regeneration of populations. First, stem cells and the field of regenerative medicine offer an alternative to the post-war social economy of tissue regeneration, the blood banks and organ donation systems that rely on gratuitous donation of tissues from one citizen to another. Since the HIV contamination scandals of the 1980s, blood donation has been in decline throughout the developed world, and blood supplies are habitually at dangerously low levels everywhere (Waldby et al., 2004; Waldby and Mitchell, 2006). Although organ donation rates have increased slightly in many developed countries over the last 10 years, the demand for organs has far outstripped supply (World Health Organization, 2003). Stem cell tissue promises to ameliorate these problems because, in theory at least, it is a self-renewing and flexible substance. As one article puts it, stem cells could act as 'universal donor cells ... "off the shelf" reagent, prepared and/or additionally engineered under good manufacturing practices readily available in limitless quantities for the acute phases of an injury or disease' (Snyder and Vescovi, 2000, p. 828). Stem cells are imagined as an unlimited resource, the precise opposite of rare, singular organs.

Second, and perhaps most significant, stem cell technology offers the possibility that the ageing body of the citizen may yet be able to extend working life and avoid long periods of nursing care at the end of life. As Cooper (2006) notes, ageing today is conceptualized in biology as an intracellular process, and stem cell lines, with their ability to replicate indefinitely, seem to overcome the problem of cell senescence (cessation of division), which afflicts cells at a certain point. The health problems associated with ageing – degenerative conditions such as Parkinson's disease, Alzheimer's disease and heart disease – are, in part at least, those of poorly regenerating tissue. These diseases are the focus of a significant amount of the global stem cell research effort, as health systems devote an increasing proportion of their budgets to the long-term management of such conditions (Chief Medical Officer's Expert Group, 2000). ESC technology offers the prospect of regenerating the tissues within ageing bodies, hence of rejuvenating ageing populations and extending the viable, non-dependent life of the population.

The ageing of populations in the developed nations also presents a major commercial incentive for stem cell research – providing potentially lucrative, rapidly expanding markets for regenerative therapies. Again, the dynamics of economic globalization contribute to the emergence of this market. The advent of neoliberal national administrations in the 1980s with tax reduction agendas, the marketization of profitable or niche sectors of health service delivery and therapeutics, and the growth of transnational health management organizations and medical tourism facilitate the growth of private or quasi-private global markets for medical therapy.

> Marketization and increasing globalization in individual sub-markets within the health-care sector are beginning to generate a range of global markets for health-related goods and services. ... First, there has been the increasing exportation [*sic*] of particular models of provision and financing, largely drawing on the private health system of the US. Examples include the expansion of Health Maintenance Organizations into predominantly middle-income markets in Latin America and ... South Africa. Secondly, we now see the emergence of global markets, where buyers and sellers ... circumvent national boundaries. With the expansion of communications technology, notably the Internet, major barriers to transferring health-related goods and services between countries are decreasing.
>
> (Kumaranayake and Lake, 2002)

Health consumers around the globe provide markets for unsubsidized 'luxury' medicine, such as cosmetic surgery and anti-ageing therapies; reproductive services such as IVF, Preimplantation Genetic Diagnosis (PGD) and sex selection or; pharmaceutical treatments for cancer and other serious conditions excluded from national pharmaceutical benefit schemes on the grounds of cost or dubious efficacy. The market for these kinds of treatments is set to expand as the populations of the developed nations grow steadily older and live longer with more chronic diseases. As Neilson (2003) notes:

> There can be little doubt that the intensity of capital investment in this sector, which now drives the economy of certain subnational regions (such as the Boston/Cambridge area in the United States), relates to the expectation of high returns as new technologies of rejuvenation become marketable to an aging population.
>
> (p. 181)

Here the biovalue generated by biotechnology translates directly into market value, both through the direct provision of for-profit therapeutics and the promissory value generated by expectations of future profits.

National policies, global pressures

Developed states have strong incentives to support regenerative medicine. A strong national stem cell research programme promises international scientific prestige and commercializable therapies that would help shore up the nation's claim as a strong competitor among knowledge economies. It gives sound inducements for high-profile scientists and laboratories to relocate from less advantageous environments, bringing venture capital with them. Successful stem cell therapies could dramatically expand the nation's repertoire of clinical treatments available to manage populations with growing proportions of the ageing, the disabled and those living with chronic degenerative conditions.

Should they choose to respond to the demands and opportunities of the new knowledge economy of stem cell science, states can make the decision to intervene at any or several points in the process of knowledge production from the basic science to the market product. Such decisions are highly political because they involve the allocation of scarce scientific resources, the acceptance of the risk involved in speculative science, the assumption that public support for the particular field can be maintained over time and, above all, the need to outmanoeuvre a state's global competitors. When analysed in terms of the knowledge-production process, each political intervention becomes a policy component that, if brought together in a sequence, constitutes a state strategy.

In deciding whether to intervene and, if so, with what policies, states face a wide range of choices (Table 0.1). They may take the view of the United Kingdom that competition for national advantage requires that policy be developed in all six arenas (UK Stem Cell Initiative, 2005a). Or they may decide this is too onerous and the cultivation of particular strengths in selected arenas is preferable. In addition, the high level of uncertainty that characterizes stem cell science and its therapeutic applications means that states are likely to have different views about the commercialization of this particular field. No accepted commercial model exists for its development upon which states can draw; no package of policy components has been established that they can apply. Indeed, the technological novelty of the field challenges the skills and inventiveness of the business community as much as those of science.

Table 0.1   States and the politics of the knowledge production process: Policy components

1. The cultural acceptability of the aims, conduct and materials of the basic science and, in the event of cultural conflict, the regulation required to ensure compatibility with the dominant social values.
2. The training, retention and, if necessary, acquisition of the scientific labour necessary for the required knowledge production to take place.
3. Investment in, and organization of, the science.
4. Ownership of the new IP: The balance to be struck between the needs of the knowledge market, the freedom of science to access research results and the cultural status of the new knowledge.
5. Stimulation of the market response through support for the venture capital function, public–private partnerships and pharma engagement.
6. The protection of citizens, consumer confidence and the integrity of the potential product through the regulation of the application and testing of the new knowledge on human subjects.

Both are aware of the value of the speculative future of regenerative medicine, yet neither can be sure of how, or whether, it will be achieved or of what kind of science–market relationship is appropriate at what point in the commercialization process.

One measure of a state's commitment to stem cell science is its investment in the basic research. In the developed world, in 2003 the United Kingdom made a £45 million allocation, and in November 2004 California passed Proposition 71 establishing a $3 billion programme in stem cell science. Meanwhile in the developing countries, China, India, South Korea and Singapore have all made an investment in the field. China's annual investment in stem cell research was recently said to be between US$4 million and 10 million with 300 researchers working in 30 separate institutions (Cookson, 2005). However, these figures are set to increase dramatically. Estimates quoted in the United Kingdom's Pattison Report suggest that over the next 5 years, China's Ministry of Science and Technology (MOST – the main source of public research funds) is expected to spend between RMB 500 million (US $63 million) and RMB 2 billion (US $0.25 billion), depending on how productive the science turns out to be (UK Stem Cell Initiative, 2005b). In January 2005 India's Department of Biotechnology (DBT) and the Indian Council of Medical Research (ICMR) announced plans for a national stem cell initiative that would prioritize research funding, focus on clinical applications and promote 'stem cell city clusters' (Padma TV, 2005). Despite (or because of) the fallout from the Hwang affair, South Korea remains firmly committed to the aggressive expansion of stem cell research and, in May 2006, allocated $454 million to the field over the next decade (Kim, 2006). Meanwhile, Singapore's vast investment in its Biomedical Sciences Initiative ($8 billion committed through to 2010) continues to act as a magnet for Western regenerative medicine scientists (Elias, 2006).

The policy choices of states are made in a context in which the science and economics of hESC science are separated by the requirements of investor confidence in the context of an uncertain market. As one market commentator observed of stem cell companies: 'Products, not science, will make these companies profitable and provide returns to investors' (Herper, 2001). And in most cases those returns are perhaps 10 or 15 years away. Unlike biotech companies that already have health products on the market generating cash flow or those with products in late-stage clinical trials, most stem cell firms are either engaged in basic research or in early-stage trials, an unattractive menu for risk-sensitive investors. For this reason it is significant, and politically logical, that venture capitalists constituted one of the biggest backers of the $16 million pro-Proposition 71 campaign, openly seeking a funding mechanism to bridge the gap between basic science and therapeutic product (Keefe, 2004). In the absence of a natural linkage between hESC science and market support, promoters of the science will pressure governments and state authorities to allow public money to assume at least some of the development risk associated with this novel science and so reassure potential investors.

One way is through international partnerships brokered by governments. In 2000, for example, the Singapore government brought together its own Life Sciences Investments Pte Ltd (LSI) and ES Cell Australia (ESCA, a private Australian investment group) to create a joint investment of $17 million in the newly formed company ES Cell International Pte Ltd. The partners aimed to develop and commercialize hESC science jointly conducted by the National University of Singapore (NUS), Monash Institute of Reproduction and Development, and the Hadassit Medical Research Services and Development in Israel (Biomed Singapore, 2000). With this type of arrangement, the political task of government moves considerably beyond one of simple investment in stem cell science and into the organization of mutually beneficial collaborations among venture capitalists, science and industry. Government's brokerage role may also involve private foundations with an interest in the field. The Juvenile Diabetes Research Foundation, for example, has set aside $6.3 million for hESC study and jointly funded a $3.1 million programme with Singapore's Biomedical Research Council

(Red Herring, 2004). Clearly, potential efficiencies can be achieved by states bringing together public and private monies in terms of both the science itself and the impact on investors' view of the science's market potential.

At the same time, nation states work within a complex web of national, regional and global forces that constrain their ability to design an enabling regulatory environment. The controversial nature of stem cell research and its electoral repercussions are only the most immediate set of constraints. The transnational organization of stem cell research and the mobility of expertise and investment present difficulties in securing stable national research cultures. Stem cell companies themselves act as transnational policy networks of influence (Gottweis, 2003; Waldby and Mitchell, 2006). The sites for negotiation and legislation over stem cell policy have multiplied and are complicated. As we noted earlier, they now include much greater input from global regulatory bodies, such as the World Trade Organization, UNESCO's International Bioethics Committee and the UN's Legal Committee. For example, at the November 2001 session of the UN Legal Committee, the Vatican observer was the sole voice arguing that the proposed international convention against human reproductive cloning should be expanded to include therapeutic cloning, an important technique in hESC science. But in a UN debate on the same issue 3 years later, in October 2004, over 60 countries led by the United States supported the Vatican position. Opposed were a group of 22 countries led by the United Kingdom and Belgium with varying degrees of commitment to hESC science. No decision was reached.

Negotiating sites also include regional bodies, such as the EU, which are playing a greater role in sponsoring debate and formulating region-wide regulatory and funding policies. These global and regional sites are formally interlinked through legally binding international agreements (e.g., the WTO's Agreement on Trade-Related Aspects of Intellectual Property Rights (TRIPS)) and link in complex ways with subnational innovation centres, like the regenerative medicine research sponsored by the California state government.

* * *

In this book we investigate the dynamic interactions between national regulatory formation and global biopolitics. We consider the underlying dynamics of discursive economies of hope, ageing populations as markets for regenerative medicine, the development of global technoscapes and high-technology knowledge economies, the process of biomedicalization, and shrinking public budgets for health care and social security. This situation places intense competitive pressure on states to fund and develop attractive climates for ESC science, which promises both to improve the health and productivity of ageing populations and to develop therapies for global health markets. We consider the development of internationally circulating arguments in favour of and in opposition to stem cell research and the various transnational bioethical spaces that have opened up to try and steer these arguments towards compromise and implementation. We consider the flow of embryonic materials from south to north and the ways these flows play into broader relations around global biopolitics. We investigate the place of transnational regulatory bodies such as the EU and the UN in organizing and modifying the international and national debates around stem cell science and ways in which national debates and policies influence each other. We identify the forces that have produced a degree of standardization in the stem cell debates and the forces of scientific, commercial and bioethical uncertainty and volatility that continue to fragment the field. We also show the emergence of new systems of knowledge appropriation through patenting and its effects on the shaping of global regenerative medicine. In carrying out this investigation, we hope to contribute to specific understandings of the global stem cell arena, but more broadly, to understandings of global regenerative medicine in the age of biotechnology.

1
Globalization, Stem Cell Markets and National Interests

Throughout the developed and developing world, states are investing public funds in basic ESC science and devising regulatory frameworks to facilitate research. In the neoliberal climate that has dominated, or at least influenced, most national administrations in the Organisation for Economic Co-operation and Development (OECD) since the 1980s, the provision of large-scale public funding for biomedical research is startling when set against stagnant or declining health budgets, the steady marketization of health services and the growing pressures on universities to find corporate research funding. Moreover, it suggests that, contrary to much of the globalization literature that emphasizes an overall decline of state power in the face of transnational communications and global markets (Ohmae, 1995; Bauman, 1998; Starnge, 2000), nation states are in fact significant actors in the biotechnology sector, particularly in the stem cell sector. This involvement, however, does not imply the existence of a nationally or locally based bioeconomy. Stem cell research and development (R&D), regenerative medicine and the bioeconomy in general are intensively globalized, through multiple levels of transnational commercial, policy and research networks. McMeekin and Green, in their introduction to a special issue of *New Genetics and Society*, which focused on the UK biotechnology industries, observe: 'The [contemporary] biotechnology sector is the first science-intensive set of industrial activities which has been truly globalized "from birth"' (2002, p. 101).

In this chapter we will investigate the role played by state institutions in the fostering of hESC research, and the life sciences more generally. Politicians and policymakers tend to foreground humanitarian rationales for supporting such research, pointing to the potential for real and significant clinical effects and applications for the ageing, the disabled and those waiting for organ and tissue transplants. Although such therapeutic aims are laudable, and dearly held by many clinicians, policymakers, patient support groups, patients and caregivers, they do not fully explain the active interest of the advanced industrial states in stem cell research. We must also consider the kind of economic productivity represented by stem cell research and the ways in which the reformulation of state-market-civil society relations in the wake of post-Fordism and globalization gives nation states strong incentives to foster nodes of the global stem cell economy within their national borders.

In what follows, we examine the kinds of pressures and constraints placed on contemporary nation states by the move to knowledge-based economies, the innovation and commercial structure of biotechnology R&D, the global marketization of health services and the ageing of populations. State interest in stem cell research is, we argue, economically driven in a broad sense, with population health benefits and clinical applications assigned a secondary consideration. As erstwhile Keynesian and corporatist welfare states such as the United Kingdom, Canada, Australia and Western Europe move closer to the status of competition states, the biotechnology area assumes more importance in economic policy. HESC research has the potential also to deliver health benefits to national populations. This potential makes hESC doubly attractive for nation states that face loss of popular support and political legitimacy, as the health and welfare of their populations become a lower priority than the fostering of global economic competitiveness. While competition states seek to roll back public welfare provision and provide residual 'workfare' benefits and services (Jessop, 2002), hESC research receives state support, we argue, because it has the potential to promote national health and relieve some of the growing pressure on aged and disability pensions through the promotion of biomedical innovation.

Biotechnology knowledge economies

ESC research and regenerative medicine are among the most recent innovations in what the OECD (1989) has described as the third wave of biotechnology development. The first two waves comprise:

> *Classical Biotechnology* ... the cumulative trial-and-error developments in the production of beer, cheese and bread, etc. as well as traditional animal and plant breeding and *Modern biotechnology* ... the science-based developments from the late 19th century that led to the modern pharmaceutical and food processing industries.
> (McMeekin and Green, 2002, p. 102)

Third-wave biotechnology is characterized by the re-engineering of microbiological processes and the multiplication of industrial/medical applications in diagnostics, industrial processing, pharmaceuticals and agriculture. This third wave began to build in the 1970s. Corporations, universities and state administrations started to respond to the restructuring of the global economy in the wake of the dramatic decline in profitability of the OECD nations' Fordist industrial base, centred on mass manufacture, Keynesian economic policy and state insulation of national economies from the fluctuations of global finance. Saturated domestic markets, growth in wages and decreasing profit from mass-manufacture techniques prompted nationally based Fordist firms to expand into foreign markets and use stateless finance capital as credit, undermining the ability of the nation state to regulate or profit from their activity (Jessop, 2002). Declining profits prompted companies to look beyond the Fordist models of mass production and consumption regulated within the space of the nation state. Companies searched for new material bases of profitability in technological innovation and more flexible forms of production. They turned to the basic scientific research carried out in universities during the 1950s and 1960s in biology, communications engineering and information sciences and began to explore the commercial applications of this research. Since the 1970s, each of these areas has been developed into generic 'technology platforms' (McMeekin and Green, 2002), which 'underpin a wide range of industrial sectors without being unique to any one' (Etzkowitz and Webster, 1995, p. 495), and form the competitive basis for a growing number of sectors.

This commercial exploitation of basic science underpins the OECD nations' transition from industrial mass-manufacture economies to knowledge economies, dependent on a highly educated workforce; technical research and innovation; and extensive cooperation among university, industrial and government sectors to pursue promising research directions and strategic advantages (Etzkowitz, 2003). In this context, state agencies increasingly play a coordinating role, for example, promoting and linking national and international research networks, and shaping research directions through public–private partnerships (Cooke, 2002).

The development of these knowledge-driven sectors of the economy has a strong speculative dimension, however, in both the scientific and fiscal sense. In 1980, the US legislature passed the Bayh-Dole Act, designed to encourage commercial enterprises to invest venture capital directly in university-based biotechnology research, via IP regimes and public–private partnership research funding. Since then, many nations within the OECD have introduced similar legislation to encourage innovation-based commercial enterprise through university–private sector links, and changes in IP regimes (Mowery and Sampat, 2004). These changes have facilitated collaboration between university basic biology laboratories and biotechnology and biomedical firms interested in commercial applications.

However, many lines of basic biology research fail to deliver usable applications or sustainable profitability. The dot.com boom and bust of the late 1990s reveals the risks and difficulties associated with developing new R&D-driven industries. This uncertainty is particularly marked in the biotechnology area, where high levels of expensive expertise are required and where development pipelines tend to be long and unpredictable. As we noted in the introduction, 20 years of genetically oriented biomedical research has failed to produce safe gene therapy applications, for example. Currently, many nations in the OECD are making substantial public investments in large population biobanks, research facilities designed to facilitate pharmacogenetic drug discovery and design, but this methodology might also prove unproductive. As Rabinow and Rose (2006) observe of genomic medicine, the translation of laboratory-based biotechnical innovation into therapeutic applications is a highly uncertain process.

> [I]t is still not clear whether the new forms of molecular and genomic knowledge are actually capable of generating the kinds of diagnostic and therapeutic tools that its advocates hope for. The stakes here are high, economically, medically and ethically. They lie in the presumed capacity of genomics to form a new 'know how' that will enable medicine to transform its basic logic from one based upon restoring the organic normativity lost in illness to one engaged in the molecular re-engineering of life itself.
>
> (Rabinow and Rose, 2006, p. 212)

Despite these difficulties and setbacks, the OECD (2004, 2006) regards the biotechnology sector as relatively immature and open to much more innovation and industrial development. Biotechnology offers the best hope for a new long wave of economic growth, the OECD argues, because it has the capacity not only to develop new specific products and processes but also to change the material base of significant sectors of the economy. A bioeconomy is a sort of regenerative economy,[1] replacing non-renewable resources such as fossil fuels and environmentally toxic chemical processes with the self-regenerating and environmentally friendly capacities of living process. The OECD explains it in the following terms.

> The bioeconomy [is] the aggregate set of economic operations in a society that use the latent value incumbent in biological products and processes to capture new growth and welfare benefits for citizens and nations. These benefits are manifest in product markets through productivity gains (agriculture, health), enhancement effects (health, nutrition) and substitution effects (environmental and industrial uses as well as energy); additional benefits derive from more eco-efficient and sustainable use of natural resources to provide goods and services to an ever growing global population. The bioeconomy is made possible by the recent and continuing surge in the scientific knowledge and technical competences that can be directed to harness biological processes for practical applications. Looking to the future, new techniques in biotechnology, genomics, genetics, and proteomics will continue to converge with other technologies resulting in potentially large-scale changes to global economies in the next thirty years.
>
> (OECD, 2006, p. 1)

As Cooper (2006) observes, regenerative medicine, including stem cell research, is a clinical analogue of this idea of a regenerative economy, promising to produce self-regenerating bodies in place of the 'spare parts' bodies of the industrial era, dependent on allogenic tissue economies that circulate surplus tissues from donor to recipient (Waldby and Mitchell, 2006). However, these clinical and industrial regenerative biotechnologies are still at the level of basic science and speculative imagination. Biotechnology research in general, and stem cell research in particular, involves a search for new biovaluable technologies (Waldby, 2002) – optimum points of microbiological leverage and productivity – that could form the basis for a whole suite of clinical applications, in the same way that the microprocessor in information and communication technologies (ICTs) has formed the basis for a wholesale transformation in communications. In part, the interest in basic stem cell science is driven by the hope that understanding the basic biology of embryology, ontogenesis and tissue generation will eventually act as a motive force[2] within the biomedical sectors, a major innovation that will enable the regeneration of parts of the human body and so transform significant aspects of clinical medicine.

Biotechnology research and the competition state

Human ESC research is still at an early stage, focused on an under-standing and technical mastery of the basic biology. There is currently no hESC technology in phase-one clinical trial, at least not in North America, Europe or Australia. We cannot rule out the possibility of hESC clinical trials being currently run in China or India. HESC tech-nologies are regarded as a risky commercial proposition compared with haematopoietic and mesenchymal stem cells because of their basic research status and the social controversies that surround them. The commercial stem cell landscape reflects this assessment. In 2006, the Biophoenix stem cell report identified 106 stem cell companies around the world, but only 11 were focused primarily on hESCs. The report estimated the worth of the whole global stem cell market to be $24.6 billion in 2005, growing to $68.9 billion in 2010, but it did not expect any hESC therapies to emerge during this period. Rather, the majority of stem cell commercialization will revolve around drug screening, cord blood banking and expansion, bone/cartilage and skin regeneration, and stem cell mobilization agents. Faced with long and highly uncer-tain outcomes, and a perception that hESC research is controversial, venture capital and stock market investment have been slow to enter the field. Only $50 million of venture capital was invested in stem cell R&D worldwide in 2004, and pharmaceutical companies are investing in the area very selectively (Biophoenix, 2006).

In this uncertain scientific, political and commercial climate, many OECD nation states, including most of the G8 nations[3] have proved willing to make substantial public investments in hESC research and to develop sympathetic regulatory systems. Of all investment in stem cell research worldwide in 2004, 75 per cent was from government sources (Biophoenix, 2006). Although much of the critical literature around biotechnology emphasizes the commodification of living processes attendant on the influx of venture capital and development of start-up biotechnology companies in the 1980s and 1990s (Gold, 1996; Andrews and Nelkin, 2001; Scheper-Hughes, 2002), public investment in stem cell research is not as anomalous as it might appear. Löfgren and Benner (2005) point out that unlike the highly competitive, design-based ICT industries, the most mature sector of the new knowledge economies, biotechnology development as a national economic activity since the 1970s has required substantial state involvement and support. They situate this observation within a broader critique of the globalization literature. Following Cerny (1997) and Jessop (2002), they observe that the reductions in government spending and state activity anticipated in much of the globalization literature have not in fact taken place. They point to a transformation in the form and orientation of the state in the wake of post-Fordism and globalization, rather than to a simple, quantitative reduction in its powers. In particular, they point towards a state deeply involved in the selective promotion and support of global industry, and often of other processes of globalization as well. Cerny (1997) describes the rapid transition from the post-war Keynesian state, whose political legitimacy derived from its ability to insulate key areas of the national economy from market forces, to what he terms the *competition state*, which encourages marketization and promotes the globalization of industries within its national borders.

> Rather than attempt to take certain economic activities *out* of the market, to 'decommodify' them as the welfare state was organized to do, the competition state has pursued *increased* marketization in order to make economic activities located within the national territory, or which otherwise contribute to national wealth, more competitive in international and transnational terms.
>
> (Cerny, 1997, p. 259; emphasis in original)

While post-war economic development in the OECD focused on national self-sufficiency across a range of industries and basic economic activities (e.g., banking and steel manufacture), the competition state selectively and intensively fosters the competitive advantage of some of its industries in the global economy. Contrary to neoliberal doctrine, Cerny argues, this transition has not involved a reduction of state activity. Instead, it has shifted state intervention from decommodifying bureaucracies to marketizing ones. In many ways, it involves the *expansion* of state activity, targeting intervention and support to particular economic sectors.

> By targeting particular sectors, supporting the development of both more flexible manufacturing systems and transnationally viable economies of scale, and assuming certain costs of adjustment, governments can alter some of the conditions which determine competitive advantage ... promoting research and development; encouraging private investment and venture capital ... often through joint public/ private ventures; developing new forms of infrastructure ... removing barriers to [labour] mobility, and the like.
>
> (Cerny, 1997, p. 264)

Löfgren and Benner (2005) identify extensive state involvement with the biotechnology industries across the OECD, with states focusing support and intervention on universities and the R&D system, configuring the IP climate, making public/private research funding available, creating systems to encourage the commercialization of basic research and fostering knowledge sharing. They note a pragmatic mix of neoliberal and coordinated approaches, with an emphasis on entrepreneurship and venture capital sitting alongside state–industry partnerships, market regulation and ethical restrictions on scientific procedures.

In the case of hESC research, we can see multiple levels of competition-state support. As a potential new core technology, stem cell R&D enjoys complex kinds of strategic and institutional support, in addition to straightforward research funding, as states attempt to secure competitive advantages in future commercial developments. One prominent form of support is state funding and guidance of a series of globally oriented stem cell research networks designed to bring together regional, national and transnational expertise. In the United Kingdom, the Medical Research Council, the peak medical research body, supports and helps organize the International Stem Cell Forum, a network that includes medical research organizations from 14 states that fund stem cell research and aims to promote global good practice and encourage bilateral collaboration and knowledge sharing.[4] The Canadian government funds a Canadian Stem Cell Network that brings together scientists, policymakers and bioethicists from across Canada to 'create a critical mass of knowledge', to be 'a catalyst for the development of new therapies', to 'promote informed debate' and to 'facilitate the transfer of technologies to the marketplace' (Canadian Stem Cell Network, 2005).

In Australia, the Victorian state government has committed funds to create a secretariat for an 'International consortium of stem cell networks', which will involve the participation of nationally or regionally based networks in Canada, Scotland, North Rhine, Norway and Israel, as well as the US National Institutes of Health and the International Society for Stem Cell Research.

In this proliferation of state-supported, globally oriented research networks, we can see the new 'governance' role played by competition states, wherein states shape policy formation within an economic or social sector, not through top-down command and control mechanisms but through the coordination of negotiations among key institutions in relatively flat, transnational networks of interest (Coleman and Perl, 1999). We can also see a quest to find the best level of scale for optimum knowledge sharing and synergies. Stem cell networks range from those focused on regional economies (such as the New South Wales stem cell network focused around Sydney, Australia, or the East-of-England stem cell network, focused around Cambridge), through the national scale (such as the Canadian example), to the transnational networks centred in various major innovation hubs like London. In a highly globalized industry like stem cell R&D, no particular level of scale has clear primacy, and research centres are likely to be involved in several networks at once, as different innovation bases compete to become primary anchor points for research sharing and coordination.

State funding and coordination initiatives are also used to encourage research partnerships between commercial firms and academic stem cell researchers working on the basic science. These initiatives are generally designed to encourage communication and collaboration between basic science biologists and biotechnology companies, collaboration that Jessop (2002) notes is encouraged even in neoliberal economies.

> It is recognised that many high-growth [new technology] sectors are so knowledge and capital intensive that their development demands extensive collaboration (especially at pre-competitive stages) among diverse interests (firms, higher education, public and private research laboratories, venture capital, public finance etc.).
>
> (Jessop, 2002, p. 127)

The UK Stem Cell Initiative, for example, based in the Department of Health, is an advisory group that brings university and private-sector stem cell researchers together to develop a 10-year vision for UK stem cell research. Its goal is 'to make the UK the most scientifically and commercially productive location for this activity over the coming decade, and which commands the support of public and private research funders, practitioners and commercial partners' (UK Stem Cell Initiative, 2005c). The UK Stem Cell Bank has also been set up in part to facilitate research cooperation and sharing between public and private researchers, both in the UK and internationally. The Singaporean government's largess in pouring public funding into Biopolis and the Biomedical Sciences Initiative is intended to recruit both academic and commercial scientific expertise to the island state and facilitate cooperation and technology transfer (Van Epps, 2006). In Australia, the National Stem Cell Centre has received substantial government funding to conduct research into four platform technologies – ESCs, adult stem cells, tissue repair technology and immune system modulation technology – and 'undertake world-class stem cell research to a stage where it can attract strong commercial interest' (Australian Stem Cell Centre, 2005). The US state-based initiatives use a similar model. The Wisconsin research program includes a coordinated suite of funding and policies designed to spread the research effort across a number of sites and biomedical disciplines and to link up medical and biology research training, basic biological research, clinical research, commercial biotechnology R&D and venture capital.[5] In each case, we can see direct state involvement in the production and diffusion of stem cell-related knowledge, and in configuring the relationship between a knowledge commons around basic stem cell research in which researchers can collaborate without commercial restraint and the commodification of stem cell knowledge in IP regimes (discussed in detail in Chapter 2).

The post-welfare state and the promotion of medical innovation

ESC technology, despite its controversial status, offers competition states other potential benefits beyond those provided by successful international economic performance. Cerny notes that a major political problem for competition states is the establishment of popular legitimacy in the wake of the deprioritization of public welfare. Broadly speaking, the social policy transformations associated with the public fiscal crises of the 1970s, the shift to post-Fordist economies and the dissemination of neoliberal 'solutions' have involved

> a shift in the focal point of party and governmental politics away from the general maximization of welfare within a nation (full employment, redistributive transfer payments and social service provision) to the promotion of enterprise, innovation and profitability in both private and public sectors.
>
> (Cerny, 1997, p. 260)

Innovation and successful international competition become surrogates for national welfare, a substitution that is neither stable nor fully accepted by populations, Cerny argues. The focus on competing in the global economy 'hinders the capacity of state institutions to embody the kind of communal solidarity or *Gemeinschaft* which gave the modern nation-state its deeper legitimacy, institutionalized power and social embeddedness' (1997, p. 251).

In the health area, the post-war welfare state derived a degree of its legitimacy from the provision of comprehensive health care and its general collective management of the populations' health risks through public expenditure. Since the late 1970s, such national health systems have been restructured. States have generally not retrenched national health or public health insurance systems, but they have nevertheless found ways to expose public systems to various forms of marketization and reduced eligibility for publicly funded treatments. This restructuring has involved the selective privatization of profitable sectors; subsidization of parallel private hospital systems; means-testing and co-payment systems for treatments; and the exercise of tight fiscal controls and rigorous forms of cost-benefit analysis and evidence-based evaluation on surgical procedures, therapies and pharmaceuticals (Saltman, 2003).

Nevertheless the share of health expenditure as a proportion of GDP has steadily grown over the last two decades in the OECD, and healthcare expenditure is under continuing upward pressure from high-technology medicine and the demand for long-term care for a steadily ageing population (Jacobzone, 2003). Simultaneously, as we discussed in the introduction, decreased fertility and the demographic shift towards an ageing population is believed to depress economic growth through increased demand on welfare and health-care provision and reduction of taxation revenues (Martins et al., 2005). One response to these pressures has been the gradual demutualization and devolution of many aspects of health care from state responsibility to individual responsibility for self-care and self-provision, through private forms of health consumption (Novas and Rose 2000; Ericson et al., 2000) and participation in global markets for private health services.

Given this configuration of tensions between the deprioritization of welfare (including health), the drive to secure globally competitive industries and the emergence of global as well as national markets for therapeutic applications, states have a strong set of incentives to invest in hESC research and regenerative medicine. If the basic science of hESC technology can be mastered and therapeutic applications commercially developed, nation states stand to reap the economic benefits associated with hosting leading-edge regenerative medicine commerce, in demand throughout global markets. It is estimated that at least 300 million people in the United States, the EU, and Japan alone could potentially benefit from stem cell therapy of some kind (Biophoenix, 2006).

Over and above this, however, states may gain certain benefits of legitimacy. As Sheila Jasanoff argues, biotechnology is increasingly caught up in the political legitimacy of nation states and their abilities to deliver care and health to their citizens.

If collective defense and welfare goals remain intransigent problems, as the 'war on terror' clearly demonstrates, then the mood of the moment seems all the more hospitable to state-supported advances in the life sciences, which promise citizens fulfillment on an intimate, personal scale, through longer, healthier, more liberated lives for themselves and, in time, their genetically tailored children. ... Advances in biological knowledge seemed to add point and meaning to the modernist project of rational, science-based problem solving at a moment when doubts about the goals and instruments of modernity were increasingly in evidence. Biological science and technology projected a confident ability to take much that is mysterious, elusive, particular, and problematic in the human condition and bring it within the realms of order, prediction, uniformity, and control. The life sciences in short presented themselves as ideal instruments to states in late-modern crises of legitimation.

(2005b, pp. 36–7)

HESC research also presents possibilities and hopes for health and welfare budget containment, an attractive prospect to states concerned with balancing budgets. As we noted in the introduction, regenerative medicine has the potential to reduce disabilities associated with ageing populations and extend working life. An OECD workshop into healthy ageing identifies leading-edge biotechnological innovation as the best means to improve aged health, not only increasing longevity but also disability-free years.

Biotechnologies are revolutionising the ageing experience by offering earlier diagnoses, new treatments such as regenerative and genetic interventions and ultimately disease prevention. ... Techniques to prevent or replace lost functions are borrowing from the body's own development processes, for example, the use of pluripotent cells for cell transplants and organ regeneration, or the use of hormone therapies for lost bone and muscle mass. ... In a 20-year time horizon, it may even be possible to address the fundamental causes of the ageing process and prevent or delay the onset of its most important diseases.

(OECD, 2003, p. 12)

Furthermore, as the rapporteur's scientific summary of the workshop notes, medical biotechnology both addresses social aspirations and reduces public health expenditure, particularly when delivered through public–private partnerships. 'Technology applied to geriatric medicine not only responds to the aspirations of the elderly by providing the best hope for preventing or reversing the diseases and disabilities of ageing, it will also reduce health-care costs' (Weksler, 2003, p. 18). The workshop also noted the keen interest of large pharmaceutical companies in the area of geriatric medicine and health delivery services.

Conclusions

It is evident then that hESC technologies represent a possible solution to a complex set of economic and welfare problems for contemporary states. Success in hESC research would give the sponsoring state a leading edge in the fostering of a new industry with a huge potential global market – the market for regenerative therapies. It would attract significant investment from the pharmaceutical and medical device sectors, and be a source of national scientific prestige and economic dynamism. At the same time, it would contribute in a highly visible way to the welfare of the population, without a concomitant expansion in public service delivery. The reduction of disease burden in the elderly and disabled would help to contain health-care costs and extend active life. In the final analysis, public investment in hESC research is so attractive for states because it expresses both concern for the well-being of populations and the drive for global economic competitiveness, without any apparent contradiction between the two aims. It can, in theory, deliver health improvement not through the Keynesian mechanisms of comprehensive provision but rather 'as a positive [externality] of enterprise, innovation and profitability' (Hay, 2004, p. 40).

Of course, it remains to be seen if stem cell research can fulfil such a promise, and the path to the delivery of clinical therapies is likely to be more circuitous than the political 'big picture' acknowledges, as considerations of clinical risk, treatment costs, intellectual property restrictions and large-scale tissue procurement come into play. In the next chapter we take up these last two points in more detail, as we consider some of the *material constraints* on hESC research, the problems presented by the scarcity and uneven distribution of the stem cell lines themselves. In many jurisdictions, laboratory researchers have very little access to viable, well-characterized stem cell lines. The availability of hESC lines is shaped by a number of factors, including rates of ethical procurement of precursor tissues (embryos and oocytes), local regulatory regimes and intellectual property configurations. In what follows, we consider the ways these factors shape the global tissue economy of hESC research, their modes of circulation, availability and scarcity.

2
Embryos, Oocytes, Cell Lines: HESC Science and the Human Tissue Market

Before we discuss the impact of globalization on hESC regulation, it is necessary to acknowledge that the biological material itself – the stem cell lines – circulates through global vectors. This chapter will focus on this material and its precursors – embryos and oocytes – and elucidate the political dynamics that shape their trajectories.

HESC lines are eminently flexible and robust entities – they can be frozen, passaged[1] and distributed worldwide without apparent lost of efficacy. They can be standardized and benchmarked in various ways to create comparability across different lines.[2] The same line can be researched in numerous laboratories, conferring the many advantages associated with scientific collaboration on a standardized object (Latour, 1987). As Waldby has argued elsewhere (Waldby and Mitchell, 2006), the hESC line is a highly *disentangled* entity compared with the IVF embryo from which it was derived. Whereas the IVF embryo is caught up in a dense web of reproduction, family relations and social controversy, a gradual process of purification – donation to clinic and then to laboratory, disaggregation, immortalization, passage – transforms the embryo into a more properly anonymized, scientific object. It becomes a neutral, validated entity, denatured of its local significance. '[hESC lines] effectively no longer have a history; they have only a future and not a past' (Glasner, 2005, p. 363). As delocalized and anonymized scientific objects, they readily circulate through the global technoscapes of contemporary scientific research, with its high premium on international collaboration and distributed research effort.

However, the form of this circulation is highly complex and provides an interesting map of commercial interests, regulatory restrictions, public opposition or support and power relations, particularly between the developed and developing worlds. In this chapter, we discuss the different global 'tissue economies' (Waldby and Mitchell, 2006) that characterize the field of hESC research and consider their implications for the politics and regulation of the area. We examine the role of stem cell lines as commodities on a global biotechnology market, and the extent to which market forces propel their forms of circulation and research/therapeutic value. We also examine public-sector involvement in their circulation, particularly the role played by the various states in promoting, managing and restricting the global flow of hESC lines, paying particular attention to stem cell banks and registries and to import/export relations between consumer and producer stem cell nations. The tissue economy that precedes hESC lines is crucial for their existence; this is the sourcing of reproductive material, gametes (sperm and oocytes) and embryos themselves, and we focus on oocyte supply and demand in relation to the burgeoning SCNT research area. A well-developed global market already exists for oocytes for reproductive purposes, a market that is open to worrying degrees of exploitation of poor women who may agree to super-ovulation and oocyte harvesting for a subsistence fee. The worldwide competition for supremacy in the stem cell field, coupled with the very large number of oocytes and embryos needed to strike a successful cell line, is bound to place even greater pressure on paid sources of reproductive material, with potentially serious risks to poor women dealing with unregulated fertility clinics.

HESC lines and biotechnology markets

As we saw in the previous chapter, hESC research has received substantial degrees of public funding from a number of states. HESC lines have a number of potential commercial possibilities, although the controversial nature of the science and the early 'basic biology' stage of the research have reduced commercial interest to a certain extent. Possibilities include

1. Commercially developed therapeutic applications: The public discourse about hESC lines emphasizes their potential for tissue repair and regeneration. However, there are currently few clinical trials for non-hematopoietic stem cells, and none for ESCs. The US company Geron had carried out toxicity trials for an hESC product to treat spinal cord injury and hopes to be in phase-one clinical trials with human subjects by 2008 (Edwards, 2007). Considerable doubt remains over clinical applications in the short to medium term, and clinicians are particularly concerned about tumor formation (Snyder

and Vescovi, 2000), the control of differentiation and the non-reversible nature of cellular transplant therapy. Untested and unproven clinical applications, however, are becoming common in poorly regulated environments like India, where government agencies are struggling to introduce bioethical and clinical trial guidelines into a field where 'no one has a clear idea of what clinical studies are being carried out where, and how they are being evaluated' (Jayaraman, 2005, p. 259).

2. Multiple research possibilities: Beyond therapeutic applications, hESC lines have a number of possibilities, with more prospects of immediate or imminent applications. They can be used as in vitro surrogates for studying the progression of clinical disease, and for drug screening assays, particular predictive toxicology testing. Several companies are currently engaged in developing such assays.

3. IP and licensing: Patents can be established in the stem cells themselves or in processes involved in their isolation or culturing. Patents on stem cell lines give patent holders temporary monopolies on the commercialization of their research, allowing them to license use of their knowledge and materials for fees and royalties.

These sets of possibilities create particular flows, vectors and bottlenecks for hESC material. Rapidly expanding global interest in the commercial possibilities of hESC lines creates huge demand for well-characterized, uncontaminated, and standardized hESC lines. About 200 hESC lines are estimated to be in existence worldwide (Biophoenix, 2006), but the majority of them are poorly characterized. Several consortia have formed to try to agree on characterization standards and to gather information about phenotypic and gene expression profiles, microbiology assessment and genetic identity analysis. We explore this development in detail in Chapter 7.

The IP landscape is particularly important in determining global flows of knowledge and materials in the stem cell area. Researchers who have established IP in their stem cell lines have a strong commercial advantage because they can control the ways other researchers use their knowledge, and they can earn revenue through licensing and material transfer agreements.

Licenses [govern] the use of patented material or technology, which typically define the scope of use and require payment of an up-front fee plus royalties from sales of any products derived from the licensed technology, and material transfer agreements [govern] the transfer of tangible research materials, with negotiable arrangements regarding the scope of research use, publications, transfer to third parties, and ownership of any technology developed.

(Murray, 2007, p. 2342)

Different approaches to IP regulation across different national innovation systems create potential barriers to transnational collaboration and differential incentive structures with significant consequences for what kinds of research gets done. Currently the United States enjoys considerable advantage in stem cell patents, particularly for hESC lines. Despite the social controversies over hESC research in the United States, the US Patent and Trademark Office (USPTO) presents few barriers to hESC patents and endorses strongly enforced patents on embryonic material. However, the European Patent Office (EPO) excludes hESCs or industrial processes arising from embryos from patentability (European Patent Office, 2002). The EPO has interpreted the *European Biotechnology Directive* to mean that hESCs should be excluded on the grounds that the patenting of embryonic material is contrary to *ordre public* and morality and violates the non-commodity status of the human body. The European Commission is keeping a watching brief on the issue, but meanwhile, researchers within the EU must apply to individual state patent offices. The UK patent office, for example, has ruled that the embryo is not itself patentable, nor are totipotent embryonic cells that could give rise to an embryo (UK Patent Office, 2003). However, pluripotent cells can form the basis for a patentable cell line, because 'they do not have the potential to develop into an entire human body'; thus, 'the commercial exploitation of inventions concerning human embryonic pluripotent stem cells would not be contrary to public policy or morality in the United Kingdom'.[3] This ruling may also have considerable implications for other methods of creating pluripotent stem cells, such as for induced pluripotent stem cells.

The United States thus holds the majority of the world's stem cell patents. Between 1980 and 2005, 2041 stem cell patent applications were filed in that country, with the vast majority filed after 2001. About 20 per cent of applications filed 2001–5 are for ESCs, and 6 of the top 30 cited stem cell patents during that period include ESC types. The University of California holds the largest number of stem cell patents, but the most frequently cited patents are held primarily by private biotechnology firms (Biophoenix, 2006).

Patents can be configured in different ways, with different degrees of exclusivity. James Boyle distinguishes between low-wall and high-wall IP.

Low-wall IP protects inventors without placing excessive costs on other researchers wanting to enter the field. For example, it facilitates the sharing of basic research materials and processes at low cost. High-wall IP treats basic materials as private property and places tight restrictions on licensees, an approach that tends to restrict access to a research field and prevent the kinds of public/private collaboration and peer-to-peer sharing of information and materials needed for basic research (Boyle, 1996).

In the US ESC field, access to lines and processes has been difficult for many academic and small biotech researchers not only because of the presidential restrictions on federally fundable cell lines but also because of high-wall IP approaches by some key players. In particular, the patent filed by James Thomson, the University of Wisconsin scientist whose laboratory first cultured hESC in 1998, exemplifies the problems of high-wall IP and the innovation bottlenecks it can create. The patent, held by the Wisconsin Alumni Research Foundation (WARF), is extremely broad, covering both the method for isolating primate and hESCs and three cell lines developed from them – neural cells, cardiomyocytes and pancreatic islet cells. A further patent claims all mesodermal, endodermal and ectodermal hESC lines, regardless of the way they were derived. This effectively gives WARF rights over virtually all available hESC lines if it chose to enforce them (Murray, 2007). In 2002, WARF entered into an exclusive agreement with the regenerative medicine company Geron.

> [The licensing agreement granted Geron] exclusive rights to develop therapeutic and diagnostic products from neural stem cells, cardiomyocytes, and pancreatic islet cells and nonexclusive rights to develop products and commercialize research products that are based on other cell types. Other companies could obtain only nonexclusive rights. WARF's agreements with academic researchers included critical limitations on the purposes for which stem-cell lines could be used and on the sharing of cells with other researchers, allowing WARF to propagate its contractual conditions throughout future commercial and academic development alike.
>
> (Murray, 2007, p. 2342)

Moreover, WARF made licensing conditions very expensive. In 2005, a commercial research licence from WARF included a $100,000 up-front fee and a $25,000 annual maintenance fee, a significant barrier to start up commercial research. By 2005, only seven commercial licences had been issued (Perrin, 2005). However, it should be noted that in 2006 a research exemption was introduced for non-profit researchers, giving them access to both Thomson's hESC lines and patent rights for a nominal material transfer fee ($500), provided that no rights flow through to commercial sponsors.

The WARF patents also applied to virtually all the cell lines available in the US National Institutes of Health (NIH) stem cell registry eligible for federal funding, placing further access restrictions on academic researchers (Rabin, 2005). The Thomson group at Wisconsin, who successfully derived induced pluripotent stem cells in November 2007, immediately applied for patents for the new method. This move, if successful, would continue WARF's patenting rights and its implications for stem cell research in the field of pluripotent stem cell research.

The extremely restrictive nature of the WARF patents has attracted extensive criticism, and in 2006, a coalition of scientist and activist groups, including the Public Patent Foundation, mounted a challenge to the patents. They requested that in the absence of an *ordre public* test in US patent law, the USPTO re-examine the patents on the grounds that they did not fulfil the requirement for novelty and non-obviousness. The Public Patent Federation argued that the WARF patent technique was the same one used to make mouse ESC lines in 1981 and that scientists at the National University of Singapore had isolated hESCs in 1994 (Tran, 2006). In January 2007, WARF removed some of the more restrictive features of its licensing agreements and lowered its fees 'in order to facilitate research' (McCook, 2007). Despite WARF's gestures, the public interest challenge was not withdrawn, and in April 2007, the USPTO issued a preliminary decision that the patent should be revoked. Unsurprisingly, WARF contested the decision, and early in 2008, the USPTO reversed its revocation, upholding the original claims.

The WARF patent saga demonstrates the complex interplay of property relations and innovation ecologies, public interest and market forces in the stem cell area. The upstream, basic biology status of hESC research means that a critical mass of key researchers still subscribe to the Mertonian norms of collegial sharing, publication and disclosure rather than the high-wall IP approach adopted by WARF and Geron. In this particular case, concern about freely circulating stem cell lines, techniques and materials has outweighed concern about shoring up strict property rights, but in the absence of a public interest test in the USPTO, nothing prevents it from granting other such broad patents. We examine the global politics of stem cell patents again in Chapter 7, when we discuss the attempts by the World Trade Organization to impose global patent standards on different national research constituencies.

State and public-sector management of hESC circulation

As we saw in previous chapter, hESC research has received considerable levels of public funding; in addition to financing, public-sector involvement often takes the form of encouraging international collaboration and knowledge exchange, particularly in the early stages of basic biology research. As Löfgren and Benner (2005) point out, despite the public sector's involvement in biotechnology development and its broad partnership with industry, it is nevertheless concerned with foiling market tendencies towards the excessive privatization of research results.

> The principal focus of OECD governments in respect of competitiveness and industrial dynamics is on universities and the R&D system, training and education, support for entrepreneurship, and the commercialization of science, arrangements to ensure the availability of finance, changes to taxation systems, and intellectual property rights. That governments also support measures to facilitate knowledge sharing and networking, contradicting the tendency towards privatization and commodification of the intellectual commons, points to one of the core dilemmas of the knowledge based economy.... Many studies demonstrate convergence across countries in respect of policy measures to foster science-based industries through a mix of element from the liberal governance/regulation model such as an emphasis on financial markets, entrepreneurship, and a preference for indirect measures – and the coordinated model, including state-industry partnerships, networking and alliances, market regulation, and ethical restrictions on science and industry.
>
> (Löfgren and Benner, 2005, p. 5)

As we saw in Chapter 1, a number of national, regional and transnational stem cell research networks have sprung up, with varying amounts of public funding to facilitate this kind of research collaboration. Laboratories across the world also work directly on particular research projects, work that often involves the sharing of stem cell lines themselves. To facilitate access to and distribution of stem cell material, many countries have set up stem cell registries that list the location and characterization of particular stem cell lines. The US NIH has a stem cell registry that lists the 22 hESC lines eligible for federal funding.[4] Canada is setting up a similar registry of lines generated in that country, requiring researchers to participate in the registry to be eligible for funding. All hESC lines generated using Canadian Institutes of

Health Research funds will be listed with the registry and made available to other researchers. The International Society for Stem Cell Research maintains a registry of cell lines ineligible for NIH funding.[5]

Public stem cell banks are a more direct intervention into the distribution of hESC lines because they manage not just information about available lines but the actual lines themselves. At the time of writing, only two such banks are operational – the UK stem cell bank (UKSCB), opened in 2004, and the Spanish national stem cell bank, opened in 2005. Several other nations have the creation of a publicly funded stem cell bank under active consideration or are in the early stages of implementation, including Singapore, Sweden, the United States, India, China and Australia.

Nationally based, publicly funded stem cell banks are emblematic of the tensions between liberal market forms of biotechnology innovation and the need for social governance. The UKSCB has been set up to improve the flow of well-characterized and quality-assured hESC lines between research laboratories and to help create a research commons of hESC lines accessible to both private- and public-sector researchers. The bank will divert what might otherwise be strictly market forms of stem cell circulation into public-sector research institutions. To a certain extent stem cell lines will be sequestered from market pricing. All depositors of cell lines must agree to certain conditions of distribution,[6] which include the following provisions:

- Cell lines must not be sold for financial gain.
- Depositors must make lines available to academic researchers with minimal constraints and conditions, and no upfront fees. Fees may be charged for commercial users.
- Public-sector researchers will pay the bank the marginal costs of supplying the lines, whereas commercial users will pay full costs.
- Neither users nor depositors may pass sample lines to third parties without the explicit approval of the bank steering committee or the Human Fertilisation and Embryology Authority (HFEA, the UK regulator for fertility medicine).
- Users of lines must deposit any further cell lines developed with the bank.

These stipulations have several important effects. First, they make the bank a source of low-cost cell lines for public-sector researchers, who might otherwise have to purchase lines at market prices. In this respect, it helps to bring stem cell lines within the budget of public-sector funding. Second, they should encourage the growth of a research ecology, whereby researchers are not deterred from sharing ideas by an overly competitive structure or excessive access costs (Boyle, 2003). They should facilitate osmosis between public-sector researchers and lower barriers between commercial and public-sector researchers to a certain extent, by giving them access to the same pool of material via a differential pricing structure.

Furthermore, the bank's mission explicitly provides social and bioethical oversight for the management of hESC lines. The bank's Steering Committee evaluates the informed consent procedures and other patient protection protocols tendered with each deposited cell line to ensure they comply with specified bioethical requirements for the protection of both donor and embryo. The bank will also ensure that stem cell lines are not created gratuitously. The problem of tissue matching means that a wide genetic array of stem cell material may be needed to match any particular patient. Some estimates suggest this would involve up to 4000 lines (Vogel, 2002, p. 1784), although more recent estimates by Cambridge University researchers (Taylor et al., 2005) found that as few as 150 stem cell lines would provide a clinically adequate tissue match for 85 per cent of the UK organ-recipient population. In either case, researchers will need access to a spectrum of cell lines. A primary function of the bank will be to act as a library of available cell lines, maximizing researchers' access to different lines and pre-empting any need for individual laboratories to produce diversity in-house. The House of Lords Select Committee's *Report on Stem Cell Research* (2002) identifies a major advantage of a stem cell bank as its ability to 'minimize the need to generate new [embryonic stem] cell lines (and consequently minimize the use of embryos for research) while not impeding scientific and medical progress' (paragraph 8.24). The embryo donor consent form stresses this aspect of the bank's function:

> Keeping cell lines in a stem cell bank that can be accessed by many scientists will help to reduce the number of embryos that are needed for research. The stem cell lines will be characterized, standardized, frozen and stored for future use in approved research projects, perhaps many years later....All scientists who want to use banked stem cell lines derived from embryos will have to seek approval from a high level Stem Cell Steering Committee ... approval will only be given if i) the research is necessary and of high quality, ii) the scientists are following UK legal and ethical guidelines, and iii) the scientists provide evidence that they have secured all essential licenses or accreditations from relevant UK ethical and regulatory authorities.[7]

These guidelines are one of the bank's major contributions to the management of the stem cell economy. They demonstrate to donors that excessive demand for embryos will be curbed and that each embryo will be highly valued. No embryonic cell line will be accepted that has not passed through the appropriate informed consent procedures. Each line will be deployed to maximize its therapeutic usefulness. Participants in focus group research endorsed these functions of the bank, stating that its greatest benefit would be the ethical oversight it provided to the uses of the cell lines (People Science and Policy, 2003).

Nationally based stem cell banks, though, also exemplify the conflict that exists between national regulatory environments and the transnational organization of stem cell research. The UKSCB was created to act as an international facility, both accepting stem cell lines from non-UK laboratories and distributing them worldwide. A long list of international lines have been accessioned by the bank, primarily from the United States and Australia.[8] The bank's focus is to secure global research benefits in the interests of British citizens rather than see them diverted to transnational health markets, as strictly commercial research might do. The UK Stem Cell Bank functions as an exogenous, global actor, positioned to marshal global flows of stem cell tissues, knowledge and expertise for national benefit as well as to facilitate international collaboration and trust relations among researchers. In this respect, the bank will reassure donors that although the cell lines created from their embryos may be distributed throughout the world, the knowledge and therapies generated will be available to fellow citizens. As an international facility with strict and highly specified standards regarding informed consent and the ethical management of stem cell lines, it also works to facilitate the 'export' of West European bioethical standards throughout the world. We explore this 'export bioethics' more thoroughly in Chapter 7.

Import/Export relations

HESC lines are a scarce resource. They are technically difficult to create; the creation of one successful line may involve large number of blastocysts. Blastocysts are themselves difficult to source in many countries with active stem cell research programmes because of social controversies over hESC research, donor preference and variable rates of assisted reproduction. The prospect of a donated embryo shortage is certainly a significant anxiety among British stem cell scientists and policymakers. In other contexts, notably the United States and Germany, regulations limit the supply of embryos or stem cell lines according to the dates they were created. President Bush's speech on 9 August 2001 promised both that federal funding would be made available for hESC research and that funding would be limited to stem cell lines created before the date of the announcement. In Germany, hESC research is restricted to imported embryonic stem cell lines created before 1 January 2002, the date of the legislation. These restrictions were made so that no embryo should die for German research (Sperling, 2004). Moreover, many countries forbid the creation of 'research' embryos, that is, embryos created not for purposes of reproduction (and then deemed spare) but, rather, embryos created solely for hESC research. The United Kingdom, Belgium, Israel, India, Sweden, China, South Korea and Singapore are the only countries at the time of writing that permit the creation of research embryos. In other cases, hESC researchers rely on embryos surplus to reproductive use in IVF treatment.

The uptake of IVF also varies from country to country. In India, for example, the emphasis on fertility and the birth of sons, combined with a very large population, produces a large IVF industry, with over 200 clinics reported to be in operation, mostly in the private sector (Kumar, 2004). The embryo itself is not considered to have high onto-logical status, so bioethical and cultural barriers to donating embryos for research are relatively low. However, ethnographic work carried out in Indian IVF clinics suggests that embryo donation for hESC research is also facilitated by the exchange of free IVF for embryo donation. The researchers note that given the social stigma of infertility in India and the high cost of treatment relative to average incomes, a couple's con-sent to donate embryos is somewhat pressured by such offers, so the ethical provenance of stem cell lines derived in India should be treated sceptically (Bharadwaj and Glasner, 2004).

Israel also has a thriving IVF industry, and hESC research remains relatively uncontroversial. Based on her extensive ethnography work in Israel, Prainsack (2005) argues that the sourcing of embryos for hESC research is relatively easy because, under Judaic law, *ex vivo* embryos are not regarded as having human ontological status. Moreover, the sense of permanent threat and fragility endemic to the Israeli state encourages pronatalism and makes extensive use of assisted reproduction socially acceptable and desirable. This sense of threat also encourages the devel-opment of biomedical technologies that support the general health and longevity of the population.

> hESC research and human cloning ... [are considered] as an inher-ent part of the imagined collective body whose survival is at stake.

Endorsing a permissive approach towards technologies that are capable of sustaining the collective body by finding new cures for the sick, or even by creating new life, is the only 'rational' solution for both decision-makers and users.

(Prainsack, 2005 p. 4)

The differential abundance or scarcity of 'spare' embryos for the derivation of stem cell lines is already resolving itself into transnational export/import relations of stem cell lines. The UKSCB role transporting stem cell lines across national borders is one instance of this.

In a more local context, some Western European countries with strong scientific cultures are constrained by restrictive embryonic research regulations. Germany is set to become a major *importer* of hESC lines. In Germany, embryos are effectively citizens and full subjects of human rights, a development closely linked to the history of Nazi medical experimentation and eugenics. The *Grundgesetz*, the German Constitution, makes the protection of the human right to dignity and life the highest duty of the state, and this protection is extended to human embryos through the 1990 Embryo Protection Act.[9] This Act falls within the criminal law and prohibits, among other things, embryo research, germ line manipulation, reproductive and therapeutic human cloning, and hybridization. However, after protracted debate and dissent within and without the German Parliament, the Stem Cell Law (*Stammzellgesetz*) was introduced in 2002, which permits the importation of hESC lines under strict conditions (Sperling, 2004). In a widely noted irony, one of the stem cell line exporters to German scientists is Israel (Sanides, 2003). The stem cells made available through the NIH registry are another major source. The German Central Ethics Commission (Zentrale Ethikkommission für Stammzellforschung) has approved 15 cell lines for importation to date.[10] France and Italy have also become importers of stem cell lines as a way of circumventing tight controls over the sourcing of IVF embryos, with eight licences granted to French researchers by May 2005 (UK Stem Cell Initiative, 2005a). Israel, on the other hand, is described by one commentator as 'one of the centres where researchers from less liberal countries go shopping' (Gross, 2003).

Oocyte donation, global markets and the politics of cloning

Most nations with hESC research programmes have some reasonably streamlined means of accessing 'spare' embryos directly within their national systems, or of importing existing hESC lines if embryos are not themselves available. However, SCNT research, permitted in the UK, Australia, Belgium, Israel, India, Sweden, China, South Korea and Singapore, requires not embryos but oocytes, unfertilized ova. As we described in the introduction, SCNT is based on the technique developed to clone mammals – somatic cell nuclear replacement; this involves creating an embryo not by the usual fusion of egg and sperm, but through the in vitro insertion of the nucleus of a cell from an adult's tissues into an oocyte, which has had its own nucleus removed to make way for the introduced nucleus. SCNT turns oocytes into a much-desired human material, and their ethical procurement has become a problematic issue for regulators, feminists and bioethicists. The problem of oocyte procurement was central to the 'Hwang affair', a political crisis in South Korea that became a global scientific scandal. We discuss this affair both in this chapter and in Chapter 4.

Laboratories in most of the countries that permit therapeutic cloning find oocytes to be a scarce resource. Oocytes are already in demand for assisted reproduction, and this demand has continued to expand as the use of IVF becomes more common. Almost all UK fertility clinics report a shortage of viable oocytes for women seeking donation (Echlin, 2005). A demand for 'research' oocytes simply places greater pressure on an already short supply of oocytes and on female reproductive biology more generally. As Brown and Webster (2004) note, female reproductive biology is increasingly used by contemporary biomedicine as a gen-erative site separate from the production of children, 'through which biological materials and information is harvested for scientific, medi-cal and commercial purposes' (Brown and Webster, 2004, p. 71). The double capacities of oocytes to produce new offspring and therapeutic stem cell lines make them highly desirable, so their demand continues to escalate.

The sheer numbers of oocytes required to mount a serious research effort further drives demand here. The discredited work of Hwang Woo-Suk in South Korea gives an indication of the ratios of oocytes that may be needed to make a viable blastocyst, and of blastocysts needed to strike a viable stem cell line. In one of his studies, 18 donors produced 242 oocytes, which in turn produced 30 blastocysts and, finally, one cell line (Hwang et al., 2004a). In more recent revelations, the Seoul University inquiry in Hwang's activities found that between November 2002 and November 2005, Hwang's laboratory used 2221 oocytes pro-duced by 119 women, an average of nearly 19 oocytes each (Steinbrook, 2006). The implications of these kinds of ratios for stem cell research and eventual therapeutic applications of hESC research are quite daunting[11] and have caused disquiet among many bioethical groups. For example, the European Group on Ethics in Science and New Technologies, which advises the European Commission, singles out this problem of ratios and the inefficiency of SCNT in their advice that therapeutic cloning should not be pursued (European Group on Ethics in Science and New Technologies, 2002a).

At the same time, oocyte supply is constrained by the recalcitrance of the material and the difficulty, time, pain and risk associated with oocyte donation. Donation involves a complex IVF procedure. In a process termed ovarian stimulation, drugs are administered to shut down the woman's normal reproductive cycle, and then other drugs are administered to stimulate the development of multiple follicles. Harvesting requires invasive surgery. The procedure involves hormone injections, surgical procedures, and general anesthesia and requires extended periods of time in clinical settings. It also carries the risk of ovarian hyperstimulation syndrome, a usually unpredictable response to ovulation induction (Steinbrook, 2006) that involves pain, abdominal inflammation, possible renal failure and infertility, venous thrombo-embolism and cardiac instability. It can be fatal. Up to 5 per cent of women in treatment develop hyperstimulation syndrome (Magnus and Cho, 2005; Delavigne and Rozenberg, 2002). Moreover, little research exists about the long-term risks of ovarian stimulation, whether there may be implications for later fertility, hormonal health or the general health of reproductive organs (Dickenson, 2006). In short, ovarian induction for both fertility treatment and donation is onerous and risky, and donors are generally already in IVF themselves (Sexton, 2005).

Different jurisdictions have adopted different procurement policies to deal with the shortage of oocytes. Australia, Canada, Singapore and most of Western Europe have not altered their approach; they rely on gift systems to solicit oocytes – women must give oocytes as an act of generosity, without payment, although most regulatory systems allow the reimbursement of expenses.[12] In this regard oocytes are treated much like solid organs. The United Kingdom, with the second-largest hESC research programme in the world, has shifted its procurement policies beyond this strict gift system in a number of ways. It permits 'egg sharing' arrangements for women in IVF who agree to donate research oocytes; that is, they receive subsidized IVF treatment in exchange for research donation.

The United Kingdom, along with China, has also moved to the use of animal oocytes for basic SCNT research, an approach widely favoured by stem cell scientists who ague this is the most ethical approach for basic research that is not intended for the development of therapies. Animal eggs can be used because the procedure does not involve the mixing of human and animal genomic DNA. Rather, it produces 'cybrids', cytoplasmic hybrids, which nevertheless combine the mitochondrial DNA from the two species. The production of cybrids will allow scientists to perfect SCNT techniques with a ready supply of oocytes. However, the genetic effects of mixing mitochondrial DNA from different species are presently unknown, and there are concerns that this may compromise some gene regulation functions (St John and Lovell-Badge, 2007). Hence, the therapeutic usefulness of cybrid-derived SCNT lines is currently questionable. Moreover, not all countries that permit SCNT also permit the manufacture of cybrids. Australia recently legalized SCNT research but explicitly prohibited the use of animal oocytes. As a consequence, SCNT research worldwide will still draw on human oocytes for the foreseeable future.

In many nations in the world, oocytes are bought and sold on an unregulated market. In the United States, gametes do not fall within the purview of the National Organ Transplant Act 1984 (Steinbrook, 2006) because they are classified as 'self-regenerating tissue' and therefore can be bought and sold. In Spain, oocyte donation does not come under the authority of organ donation legislation. These countries now have a vigorous and privately controlled internal trade in reproductive oocytes, in each case linked to unregulated transnational trade. Nations that attempt to protect oocyte donation from market forces find that nationally based regulations are being increasingly undermined by tissue trading *between* states, facilitated by medical tourism, global medical commerce and the ever-expanding demand for oocytes.

This global market has serious implications for women who are not well protected by legal structures, bioethical regulations, adequate income or a feminist-influenced civil society. The Hwang case is telling in this respect. It has emerged that some of the oocyte donors for his studies were young research staff in his own laboratory, with all the implications of coercion and absence of meaningful informed consent this entails. Hwang's laboratory also used numerous paid oocyte suppliers (Steinbrook, 2006), which also raises the spectre of informed consent and the problem of full disclosure of risks to donors when demand for donated tissue is high. Currently, a coalition of 35 women's groups is involved in a suit for compensation against the South Korean government on behalf of the 20 per cent of the women who provided eggs on the grounds that they had not been informed of the risks of donation. In some cases, the women required hospitalization because of the side effects of ovarian hyperstimulation (Hwa-Young, 2006).

Next we will focus on two different kinds of oocytes market – the medical tourism markets in Spain and South Korea and the highly stratified, predominantly internal market in the United States. We will consider the possible effects that the new demand for research oocytes is having on such markets and examine the place that oocyte vendors (i.e., women who *sell* oocytes rather than donors, who *give* them) occupy in the global knowledge economies, and what this suggests in terms of improving regulation and protection for such women.

Global oocyte markets

Reproductive tourism

Shortages of gametes and regulatory restrictions have created a market for reproductive oocytes among wealthy citizens from countries in northwestern Europe. To meet this demand privately run fertility clinics have been set up on the fringes of Western Europe in countries with more permissive regulations. Clinics in Spain and Crete offer 'IVF holidays' to attract wealthy north European IVF tourists who have not been able to obtain satisfactory treatment at home. Multilingual websites and Internet communication have made international partnerships and patient bookings much easier, facilitating the growth of such IVF tourism.

Spain is a particularly attractive destination for European IVF tourists because it combines high medical standards with minimal regulation, a liberal approach that has its origins in the post-Franco government's desire to remove state restrictions on reproduction.[13] According to a report about UK IVF tourism in *The Observer* newspaper (France, 2006), clinics recruit through beauty parlors, supermarkets, colleges and by word of mouth, and pay oocyte suppliers about £1000 per procedure, with a premium paid to fair 'northern-looking' donors. The Spanish Society for Fertility estimates that approximately 3000 oocytes were traded in this way in 2004 (European Commission, 2006, p. 9).

Another recent investigation by the UK *Observer* (Barnett and Smith, 2006) found that fertility clinics in the Ukraine and other parts of the former Soviet Union recruit young East European women and send them to clinics in Spain and other locations – Cyprus and Belize, for example – to provide oocytes for north European couples. The women interviewed reported being paid between £300 and £600 per procedure, with a higher fee if they produce more oocytes per procedure. They also referred to friends who had donated multiple times. One informant, a nurse working in the industry, 'told *The Observer* that some women viewed egg donation as their main source of income, going through the process of being injected with hormones at least five times a year' (Barnett and Smith, 2006). Some also reported combined oocyte vending with a stint of work in the local sex industry.

A similar IVF tourism market was operating between Japan and South Korea at the time of the Hwang scandal, primarily through an oocyte brokerage company called DNA-BANK, based in Seoul and Tokyo. The Korean clinic recruited young Korean oocyte vendors, primarily college students with tuition fees or debts to repay, for Japanese couples who were forbidden by law to find oocyte donors in Japan (Paik, 2006). Rather than offering direct fertility services, it acted as a brokerage firm, working with several South Korean fertility clinics and matching purchasers with vendors.

The South Korean case is notable because it emerged that Hwang had used the DNA-BANK to obtain research oocytes for his discredited SCNT work, and in fact that the brokerage service had taken the initiative, offering oocytes to his laboratory. Paik summarizes the significance of this transition from reproductive to research oocyte procurement.

> The network for trafficking ova was first established to trade ova for the IVF purpose [*sic*], but, then, was utilized for trading ova for the biomedical research. In this sense, at least in the Korean case, for the ova brokers there existed only a very fine, and sometimes arbitrary, distinction between ova for research and ova for infertility treatment.... While people find the trafficking of ova for the IVF purpose harder to oppose, it is the very same network that enables the distribution of ova for research.
>
> (Paik, 2006, pp. 4–5)

The DNA-BANK case demonstrates the ease with which recruitment networks and vendor populations for reproductive oocytes can be used for research oocytes.

The US oocyte market: Niche markets, stratification and the stem cell industries

The United States has the most well-developed and least regulated internal market for oocytes. It also has the greatest number of stem cell companies, and privately funded stem cell research is unregulated at a federal level. Reproductive oocyte trading is routine. The US Centres for Disease Control and Prevention report that in 2002 alone, purchased oocytes were used in 13,183 (11.4 per cent) of the 115,392 procedures involving assisted reproductive technology, for fees of around $4000–$5000 per cycle (Steinbrook, 2006, p. 324). Like the European market, the US reproductive oocyte market is stratified according to the appearance and 'racial' characteristics of the vendor. Premiums are paid for vendors with additional desirable characteristics, particularly pretty, athletic women at elite colleges, who are routinely offered very high fees. While it is difficult to get verifiable, up-to-date figures, fees are reportedly between $20,000 and $100,000 per cycle (Pollock, 2003).

There are no federal regulatory barriers to prevent the reproductive market in oocytes becoming a market for research oocytes as well. US bioethicist Jeffrey Kahn argues that the general US aversion to public funding for hESC research makes oocyte and embryo purchasing on an open market more likely because funding agencies provide no funding for embryo collection, even in public-funding situations.

> Because of sensitivity over the status of human embryos and federal law that prohibits tax dollars from being used for embryo research, the U.S. National Institutes of Health (NIH) has proposed that it will fund research on stem cells but won't fund the collection of the stem cells themselves. This leaves private companies to act as suppliers of stem cells. Where will the embryos come from, what limits should there be on embryo use, and how close are we to a market in human embryos? ... The government is effectively the market maker – a public buyer creating a demand to be filled by private suppliers.
>
> (Kahn, 2000, pp. 1–2)

It is currently illegal for couples or women to sell their spare IVF embryos, however, so private suppliers would be likely to target financial incentives at oocyte providers. Although reproductive oocytes provided by fair-skinned college students fetch high prices, phenotype is irrelevant in SCNT research. There is considerable scope to extend research oocyte markets to poor, uneducated, and dark-skinned women, who would not normally be valued in the reproductive market, except as gestational surrogates (Pollack, 2003). In the United States, the juxtaposition of poor, ghettoized populations with high technology corridors – for example around Bethesda, Boston, Raleigh-Durham and Southern California – make these kinds of markets even more feasible. Here we can see an internal version of the extra-territorial oocyte trade already described, with poor female populations within the nation state acting as potential vendors for national biotechnology industries.

One business model for this kind of enterprise is the Bedford Stem Cell Research Foundation, founded in 1996 in the Boston area. The foundation claims to be the first organization in the world to solicit women to 'donate' oocytes purely for research. Since 2000, it has recruited oocyte vendors from the Boston area via newspaper advertisements, paying them about $4000 per procedure. The foundation conducts research within its own laboratories, supplies research oocytes to Advanced Cell Technology and is set to supply other Boston area researchers. The website (www.bedfordresearch.org) emphasizes the use of 'mild hormone stimulation' to avoid hyperstimulation syndrome and the generally high level of screening, informed consent and ongoing care provided for oocyte suppliers. The foundation only accepts women between the ages of 21 and 35, and they must already have at least one child, as a demonstration of viable fertility. According to an interview with director Anne Kiessling, 391 women by the end of 2005 had inquired about the programme. After screening, 28 started hormone injections, and 23 completed the process. Eight of those 23 donated twice; three donated three times. The donations yielded 274 oocytes, at an average cost of $3673 per egg, once screening costs were factored in (Vogel, 2006). No independent assessment of this programme was available at the time of writing. However, what is striking about its approach is the evident concern to avoid possible objections about the exploitation of its vendors and to underwrite the ethical provenance of the oocytes within the terms of the US free market.

Conclusions

Two major dynamics shape the global circulation of stem cell biological material. One is the tension between proprietary innovation and public domain research. High-wall patents at the early stages of regenerative medicine innovation present barriers to public-sector researchers and to collaboration between private- and public-sector actors. Such barriers have a dampening effect on the early stage, basic biological research and large-scale collaborative effort necessary to translate biological theory into clinical and laboratory applications. Hence much state intervention, particularly around registries and stem cell banks, is designed to mitigate the effects of excessive privatization and premature commercialization. In doing so, the regenerative medicine field is following a familiar pattern of public/private compromise. Like the Human Genome Project (HGP), many regulators, funding bodies and scientists are committed to keeping the basic biology of hESC in the public domain and, like the HGP, this compromise is imperfect and open to contestation. The WARF patent is a case in point, in which patent claims that many regard as excessive are under challenge from civil society action by citizen-scientists and social activists.

The second key dynamic is between supply and demand for various biological materials, notably embryos and oocytes. Different national systems evince dramatically different abilities to ethically source these materials so that some can afford to export while some are forced to import. At present, the ethical sourcing of embryos is reasonably well embedded in the IVF clinics and regulatory systems of many stem cell research nations, whereas the ethical sourcing of oocytes is much more volatile, both because it is in the early stages of regulation and because of the existing unregulated global market for reproductive oocytes. At the same time, the worldwide scandals over Hwang's questionable methods of oocyte acquisition has alerted regulators, stem cell research companies and professional bodies to the potentially devastating effects of bioethical issues on research reputation.

The situation in the United States is particularly volatile. The dramatic inflation of prices around reproductive oocytes over the last 5 years and the expansion of markets to include research oocytes have sparked concern among many key actors, and scientific peak bodies are working to eliminate the unregulated trade in oocytes. In 2005, the US National Academies of Science recommended in its 'Guidelines for Human Embryonic Stem Cell Research' that no payments should be provided for donating gametes for research (Steinbrook, 2006). Some liberal supporters of stem cell research, in assessing the Californian Stem Cell Initiative, argue for 'public sector bodies with the power to establish and enforce comprehensive regulations that apply to both publicly and privately funded research' (Reynolds et al. 2006, p. 17). They advocate adequate reimbursement rather than payment, an institutional separation between oocyte harvesting clinics and stem cell research companies, and an ongoing duty of care to donors, with adequate follow-up and research of long-term consequences. The California Institute for Regenerative Medicine, established in response to the 2005 California stem cell initiative, prohibits payment for oocytes, although it permits compensation (Steinbrook, 2006), aligning it more closely with West European and Commonwealth norms, and presumably smoothing the way for international collaboration.

So, both key dynamics, public/private tensions and the political economy of supply and demand, demonstrate what is at stake in regulation of the biological material itself. Regulation shapes who donates or sells material, the conditions of its sourcing and its relative abundance or scarcity. It shapes who has access to research material, who can profit from it and the boundaries between private and public equity in research outcomes. The Hwang affair demonstrates what can go wrong if the sourcing of materials is not properly managed and brought within a bioethical purview. The controversies over the WARF patent indicate the social tensions that can develop if material circulates only according to market logics, without other values being considered.

3
Global Regulation and Local Policy Narratives: Making Sense of Dolly

In the last two chapters we described the rise and development of the increasingly differentiated and complex global stem cell economy. Closely related to the operation of the emerging stem cell economy, and regenerative medicine more generally, is a regulatory regime that deals with the multiple societal, legal and ethical challenges in this field. The globalization of stem cell research has heightened pressure to build a regime of regulation that complements globalizing trends in the bioeconomy. But as this and the following chapters will show, a global regime of regulation must deal with the persistence of local policy narratives that until today has led to striking regulatory differences among nations. A central characteristic *and* challenge of regenerative medicine is the striking contrast in how its projects, ambitions, goals and strategies are perceived and constructed in different countries. In some countries (e.g., the United Kingdom and South Korea), hESC and SCNT research are permitted under certain regulations, but in others (e.g., Germany, Japan and Italy), such research is completely or partially prohibited. We will probe the questions of why such differences exist, whether any signs for a rapprochement between countries are present and what the political implications are for these differences between nations in research and regulation.

Setting the stage for hESC policymaking

In the following two chapters we will take a closer look at the political dynamics that have led to the emergence of significantly different regulatory regimes across the globe. Few countries in the world forbid hESC research completely, but there are considerable differences in the types of hESC research experiments that can be done and under which regulatory conditions. Scientists and patient groups complain that these differences create environments in which hESC research can flourish in some countries, while in others just the opposite is the case. HESCs retain the same qualities if they are used for research in Berlin, Seoul or London, but apparently, citizens, policymakers, scientists and journalists in these places perceive this research in different ways and draw very different conclusions as to what is acceptable research, and what is not. We will argue that the substantial differences between countries in their approaches towards stem cell regulation can only be partially explained by reference to culture, religion or general path dependency. While these factors play a role, others need attention.

The social debate about hESC research was by no means a 'sudden event' and cannot be conceptually separated from other fields of social debate dealing with the question of the boundaries of human life, such as abortion, embryo research and animal cloning. We will argue that thematizing the boundaries of human life carries with it an inherent socio-political explosiveness that depends greatly on interpretations of what is at stake in a particular policy field and what kind of political action is required. Political reactions to such developments significantly depend on narrativizing these developments and shaping different policy dramaturgies with resulting impact on the emergence of regulatory regimes. It is during these 'stem cell dramas' that not only ordinary people but also policymakers and journalists make sense of stem cells and structure their preferences and options for regulatory policy action.

When in late 1998 hESC was breaking news, the first reports in the mass media described these successful experiments as a 'medical breakthrough' and 'scientific revolution'. But soon other aspects of stem cell research gained attention, such as its 'ethical' implications and the possible need to devise strategies for its regulation. Within months, hESC became a highly controversial topic, alternatively connected to such diverse themes as medical progress, magic cures, the killing of babies and moral decline. The meaning of 'human embryonic stem cell research' was not just controlled by scientific papers or press conferences by medical researchers presenting their work but determined by numerous influences and suggestions of interpretation offered by a variety of actors, ranging from scientists to medical doctors, patient groups and 'pro-life' advocates. These various sources of interpretation simultaneously created a new discourse on 'human embryonic stem cells' and connected stem cell research to earlier statements and discourses such as health policy or abortion, something we will call *discursive memory*. It is this 'pre-history' or historical context of the stem cell and cloning research debates that tended to be mobilized in many countries – a story full of ethical considerations, emotions and uncertainty – but it is also a story that, in numerous ways, is strikingly different from the cloning and stem cell discussion.

Discursive memory (Maingueneau, 1984, p. 131), memories of earlier controversies, interpretations and assessments of 'dealing with embryos and fetuses', became an inseparable part of the hESC research debate and played a crucial role in shaping the new controversy. Spaces of policymaking usually have some sort of history in which past and present are linked by discursive memory, and this played an important role in the shaping of public policies dealing with stem cell research. Discursive memory is selectively mobilized in the form of policy narratives. Narratives referring to the 'pre-history' of a policy are typically contingent and should be regarded as material actors who mobilize under certain circumstances to structure a space of policymaking. In this respect, the abortion and embryology/fetal research debates from the late 1960s until the 1980s were a prelude to the cloning and stem cell controversy, but they were not simply a continuation of one story.

In the United States, Germany, the United Kingdom and in Europe in general, the creation of the scientific-political field of hESC was strongly connected to the discursive memory of the abortion debates, discussions about advances in reproductive medicine and earlier struggles over fetal and embryo research. This discursive memory materialized in a complicated structure made up of institutions, laws, regulations, practices and argumentative strategies and repertoires that shaped the stem cell controversy in a significant manner from the very beginning. Both opponents and supporters of hESC research attempted to interpret this new line of research as either a reflection of the earlier controversies or as a new phenomenon altogether, which had to be dealt with in new ways. At the same time, existing laws or court traditions were elements of a complicated topography within which stem cell research came to be located.

There were also significant differences between the debates in the late 1990s and those in earlier decades. Neither the abortion issue nor embryology and fetal research were closely tied to narratives of economic and industrial relevance. Moreover, embryo and fetal research in the 1970s and 1980s were usually depicted as much further from application than was the case with stem cell and cloning research.

Narratives and discursive memory organize the way we think about something like stem cell research; thus they play a key role in shaping policy dramas and the staging of a particular policy controversy. A *policy dramaturgy or scenography* comes into existence through a specific kind of discourse or mode of argumentation that defines the roles of the actors in a policy field and their identity, a topography, a location where a process of policymaking takes place, and a chronology, a time sequence when policymaking process takes place. Policy scenographies determine possible modes or argumentation, what can be said and what cannot be said. Thus, policy dramaturgy can be seen as the strategic moment in a regulatory process, a moment that is not simply up to the decisions of individual actors but operates in a complex discursive context. Regulatory policymaking is a multifaceted process that involves argumentation as a process of shaping and creating a dynamic, a rationality, a logic of reasoning, a basis for decision-making. The way a certain policy problem is depicted and defined gives rise to particular scenarios of interaction and involvement, describes involved actors, a particular timing, and the location for a policy development to take place.

In the following sections, we follow in detail the shaping of different policy scenographies in hESC research, comparing two divergent regulatory models, the United Kingdom and the United States, and complementing this analysis with further case studies on Germany, Italy, Japan and South Korea. Whereas hESC policy in the United Kingdom is a good example of adopting a liberal approach towards stem cell research, United States' federal policies offer an example of a restrictive approach. The case of Italy is interesting because it raises the question of how one of the most permissive regulatory regimes could, almost overnight, turn into one of the most restrictive. Finally, South Korea and Japan are two examples of non-Christian countries, for which the ethical implications of hESC research had considerable impact on policymaking, and have led, in the case of Japan, to relatively restrictive legislation.

To understand the emerging regimes of stem cell governance since the late 1990s one must, first of all, see them as a project of 'dealing with Dolly' (Franklin, 2007). The birth of Dolly, the sheep, was the key event in the shaping of the local regulatory conflicts around hESC and cloning research across all the countries we discuss in this book and that we believe are currently the key actors in global stem cell research. Dolly's birth constituted what Laclau describes as a *dislocatory moment*: 'the emergence of an event, or a set of events, that cannot be represented, symbolized, or in other ways domesticated by the [dominant] discursive structure – which therefore is disrupted' (Laclau, 1990, p. 41). The dislocatory moment not only refers to a 'traumatic event' of chaos or crisis (Torfing, 1999, p. 149) that induces a break with dominant structure but also denotes a turning point or rupture of a discursive structure, calling for a process of policy reordering. As we argue, the somehow 'sudden' availability of cloning technology with the potential to 'replicate' complex organisms constituted just such a rupture of pre-established modes in a particular policy field. Through 'Dolly', the exact delineation of the boundaries of life came into question, the human potential to 'create' life, the 'beginning of life' and the termination of life. To develop political decisions under such conditions of radical uncertainty became a dominant challenge for policymakers and resulted in the shaping of starkly different regulatory regimes. The political need to react to Dolly was given by the urgency to simultaneously interpret or frame 'cloning' – and to propose an agenda for political action. Thus, 'dealing with Dolly' involved a number of strategic options and choices and considerable definitional work.

The first question in our comparative analysis is whether there was any more or less coherent governmental reaction to the new availability of cloning technologies. Such a reaction potentially involves efforts of interpreting and framing Dolly, and suggests a course for political action and the shaping of a policy scenography. We argue that there were two basic models to deal with the destabilization of biomedical and regen-erative medicine discourse caused by Dolly: a *coherent*, non-polarizing approach that attempted to involve all the relevant actors in a broadly accepted, homogeneous narrative and an encompassing system of regulation for hESC research and SCNT (United Kingdom, Japan); and a *heterogeneous* approach characterized by a highly antagonistic political constellation (United States, Germany, Italy) or the repression of politi-cal debate (South Korea), leading up to missing or delayed regulation. Each scenario involved highly specific ways to 'deal with Dolly' and with it boundary of life issues, to define the nature of the policy chal-lenge, the involved policy actors and the appropriate location for the regulatory conflict to be negotiated.

Furthermore, the adopted governance strategies to deal with cloning and stem cell research implied different ways to deal with trust in the political process. Whereas in the United Kingdom and Japan, there was much emphasis on the generation of trust in the development of regulatory structures, in the United States, Germany, South Korea and Italy, the delayed and heterogeneous strategies to deal with cloning and stem cell research resulted in a crisis of trust in the developing regula-tory institutions.

Why was cloning important in the context of the global regulation of stem cell research? First, unlike hESC research, cloning research had been broadly debated in all of the countries discussed in this book.

Even countries with very liberal regulations in the field of hESC and SCNT research had at some point in time broad social debate on cloning technology. Similar to the impact of IVF technology and the legality of abortion on the social imaginary, cloning seemed to open up unprecedented and highly controversial scenarios for human intervention and thus required quick political action on the highest level.

British Dolly

Just as in most other countries with a strong interest in biomedical research, the birth of Dolly, the sheep, created a period of uncertainty and questioning in the United Kingdom. The British public and the mass media did not greet the production of Dolly the (Scottish) lamb with unequivocal enthusiasm. Calls for a ban on 'human cloning' were strong in the United Kingdom (Butler and Wadman, 1997, pp. 8–9; Lee, 1999, p. 56), and the government felt a 'need for action'. In the United Kingdom, unlike in Italy and the United States, the government reacted decisively to the 'dislocatory challenge' of the birth of Dolly.

Unlike the United States, Italy or Germany, the United Kingdom was well prepared for the upcoming hESC debate through a carefully constructed regulatory framework dealing with embryo research and artificial reproduction during the 1980s. As we shall see, the mobilization of the discursive memory connected to these debates played an important role in shaping British regulations. After the extensive social debate on abortion during the 1960s, an advisory group was appointed in 1970 under the leadership of Sir John Peel to consider the ethical, medical, social and legal implications of using fetuses and fetal material for research. The Peel Report, published in 1972, recommended that research should be allowed on fetuses weighing less than 300 grams, provided there was no objection from the mother and the research had been approved by an ethical committee. This report, which remained mostly uncontested, paved the way for a calm period of early fetal research throughout the 1970s.

However, soon after Margaret Thatcher took power, the issue of medically assisted reproduction moved to the centre of political interest, and in 1982 a 'Committee of Inquiry into the Human Fertilisation and Embryology' formed with the task

> to consider recent and potential development in medicine and science related to human fertilisation and embryology; to consider what policies and safeguards should be applied, including

consideration of the social, ethical and legal implications of these developments, and to make recommendations.

(Warnock Report, 1985, p. 4)

The creation of the Warnock Committee marked the beginning of an eight-year-long struggle about the regulation of embryo research and medically assisted procreation, a struggle that resulted in the HFEA of 1990 , the kind of reproductive medicine regulatory framework that never developed in the United States and Germany. At the core of the HFEA was the Human Fertilization and Embryology Authority, a new agency in charge of granting licences for research and treatment, licences for treatment, licences for storage and licences for research. In Germany and in the United States, many of the medical developments in the field of assisted-medical procreation and embryo research remained either unregulated or only partially regulated and, to some extent, were in conflict with the liberalization of abortion laws. In Great Britain, the HFEA constituted a pragmatic step from the abortion reforms of the late 1960s to developments in embryo research and medically assisted procreation during the 1980s. In the United States the question of the definition of human life was left to the pro-life groups, and in Germany it was defined as demanded by pro-life groups, but in the United Kingdom the issue of the definition of the embryo occupied a prominent space. In its 64 recommendations, the Warnock Committee proposed an independent body answerable to Parliament and narrowly restricted by the law (Mulkay, 1997, pp. 16–17), but it also decided to allow embryo research up to the fourteenth day after fertilization. As the report explained:

> The question was not, as is often suggested, whether the embryo was alive and human, or whether, if implanted, it might eventually become a full human being. We conceded that all these things were true. We nevertheless argued that, in practical terms, a collection of four or sixteen cells was so different from a full human being, from a new human baby or a fully formed human foetus, that it might quite legitimately be treated differently.
>
> (Warnock Report, 1985, p. xv)

The proposals regarding human embryos were cautious; for example, the creation and manipulation of IVF embryos, it was suggested by the Warnock Report, should become a criminal offence unless licensed by the proposed statutory authority. However, the reception of the report and the passing of the recommendations produced considerable conflict.

Faced with a conservative majority in Parliament and the effort by Society for the Protection of Unborn Children (SPUC) and LIFE not only to block embryo research but also to reconsider the 1967 abortion legislation, it took almost 8 years until, essentially, the main recommendations of the Warnock Report led to the passing of HFE Act (Betta, 1995, p. 53).

In the late 1980s, fetal research regained prominence as medical research began to expand the scope of its possible applications to a multitude of diseases and treatments, particularly to the treatment of Parkinson's using human fetal tissue for brain grafts. The British government invited a second committee, led by Reverend Dr John Polkinghorne to review the Peel Committee's guidelines (Polkinghorne Report, 1989). A principal concern was that women might be pressured into terminating a pregnancy to provide fetal material. The new guidelines for research accepted by the Department of Health in 1990 provided, among other things, for a total separation between the performing of abortions and fetal research, and their guidelines coincided with the passage of the 1990 HFE Act. As a result, from 1990 on the United Kingdom had a comprehensive regulatory framework for reproductive medicine, able to deal with issues ranging from medically assisted procreation to embryo and fetal research. While the Warnock Committee's decision to allow embryo research up to the fourteenth day after fertilization remained contested, research was allowed to continue within the framework of a tightly structured regulatory system.

From Warnock to Dolly

When Dolly arrived at the doorstep of UK politics, the questions of hESC research and embryo research already had a long history and an institutional framework that potentially could take care of emerging regulatory issues. However, it was unclear whether the HFE Act actually covered the newly available cloning techniques, so an inquiry by the House of Commons Science and Technology Committee was conducted and the findings published in March 1997 in 'The Cloning of Animals from Adult Cells' (Science and Technology Committee, 1996–7). The Committee argued that concerns over the cloning of Dolly may have overshadowed potential benefits. It suggested that the Human Genetics Advisory Commission (HGAC) would advise on the implications of the work for human genetics and also recommended that work that would create 'experimental human beings' should not be carried out, suggesting that Parliament should reaffirm a ban on human reproductive cloning. The HGAC was established in December 1996 as a non-statutory advisory body that provides independent advice to UK Health and Industry ministers on issues arising from developments in human genetics that have social, ethical or economic consequences. The commission is charged with setting its own priorities, although from time to time it may be requested to provide urgent advice to ministers. The commission was also asked to advise on ways to build public understanding of the new genetics. The HGAC and the HFEA decided to explore ways of holding a public consultation exercise into the implications of cloning developments. A joint working group, consisting of members of both bodies, was established to consider the planning, drafting, distribution and analysis of a joint consultation paper on the issues for human genetics arising from advances in cloning technology.

The government response to the Science and Technology Committee's report was published in December 1997 (Government, UK, 1996–7). It reiterated the Minister of State for Public Health's statement, explained that the HGAC and the HFEA were exploring ways of holding a public consultation exercise on cloning and said that the government would consider carefully, in the light of developments, whether the legislation needed to be strengthened in any more specific way, taking into account the views of Members of Parliament, the HGAC, the HFEA and responses received to a more general consultation on the broader issues (HGAC, 1998a, 1998b).

The consultation document, 'Cloning Issues in Reproduction, Science and Medicine', attracted substantial interest. More than 1000 copies were distributed, and the document was also distributed through the HGAC website. The consultation sought general comments about how the technology that led to Dolly might develop and the opportunities and problems that might be raised by human reproductive cloning and other applications of SCNT technology. It also invited views on priorities for the future and the ethical settings in which these scientific developments are taking place. It sought comments on any other ethical issues raised by human cloning that respondents identified. It was requested that responses be structured around replies to six questions. Respondents were also invited to make suggestions about what advice might be offered to ministers on ways to build public confidence and understanding of the new developments in genetic techniques.

The document drew, apparently for the first time, a distinction between two types of cloning:

> For the purposes of this consultation we draw the distinction between two types of cloning: on the one hand, human reproductive cloning,

where the intention is to produce identical foetuses or babies; and, on the other hand, what may broadly be called therapeutic cloning, which (although not coterminous with conventional scientific usage) includes other scientific and medical applications of nuclear replacement technology.

(HGAC, 1998a)

This distinction between 'therapeutic' and 'reproductive' was not based on any pre-existing scientific terminology. It was developed by the Committee and eventually, shaped in a most powerful way the discussion about the topic of stem cell research and cloning in many countries (Interview, HFEA/Oxford University, Oxford, 5 July 2002).

The consultation paper raised many fundamental questions concerning cloning, such as whether any areas of medical research could benefit from cloning research. But the paper also began to outline a particular policy direction. On the one hand, it related what it defined as *therapeutic cloning* to the current regulatory practice:

The HFEA's policy is that it will not license any research which has reproductive cloning as its aim. However, it would consider licence applications for other types of research involving embryo splitting or nuclear replacement in eggs, provided that the research falls within one of the purposes specified by the HFE Act, or any regulations, which may be made by the Secretary of State for Health as described above.

On the other hand, the report clearly highlighted the possible advantages of human cloning in a range of applications. Nearly 200 responses were received – about 40 per cent from individuals and the rest from a wide range of constituencies – scientists, clinicians, academics, religious groups, ethicists, lawyers, industry and lay groups.

The final December 1998 HGAC paper spoke a clear language and was a strong endorsement for opening up the door for hESC research. Despite this clear endorsement of support for therapeutic cloning as a field of vital politics by the joint HGAC/HFEA paper, it was by no ways clear that this position would become official government policy. In fact, the BSE scandal and the broad discussion about genetically modified foods had created a discursive context for the shaping of governmental strategies in other sensitive areas of policymaking. In June 1999 the government decided to defer its decision on a possible rewriting of the HFEA regulations as suggested in the joint HGAC/HFEA report and suggestions.

When the HGAC/HFEA report was drafted, it mainly had focused on cloning issues, but the work by James Thomson and John Gearhart in human stem cell research had not yet been published, and thus it was difficult to take these developments into account. The public health minister, Tessa Jowell, told Parliament: 'We believe that more evidence is required of the need for such research, its potential benefits and risks, and that account should be taken of alternative approaches that might achieve the same ends' (quoted in *BioCentury*, 1999). It was announced that a new expert group would look into the issue of therapeutic applications of stem cell research and cloning. As a high level Department of Health officer described the results of the HGAC/HFEA report:

> I think government at that time felt that was the right way forward, but that it recognized, that it had come to that conclusion without having necessarily carried out a very substantial analysis to where the science was taking us....The HGAC/HFEA report was already in press when J. Thomson first isolated stem cells, so almost immediately there was this new area that the report had not covered, so that let the government say, we think this is the right way forward but we really want to put together an expert group, scientists, medical and academic scientists, ethicists, lawyers, specifically to look at the implications of cell nuclear replacement research and stem cell research.
> (Interview, Department of Health, London, 2 July 2002)

Immediately following this move of the government were threats from the biotechnology industry. Simon Best, the managing director of Geron BioMed, the company with exclusive licence to commercialize the technology that created Dolly the cloned sheep, said his company had hoped to collaborate with other institutes in the United Kingdom, but it was looking less likely now and Geron BioMed might conduct its human embryo research in the United States. The UK Bioindustry Association noted that other countries such as France, Germany and Spain were reconsidering bans previously implemented on the therapeutic use of human embryo cloning, and that the money for research would move elsewhere if research were not allowed in Britain (*BioCentury*, 1999; *Independent*, 1999).

It seems that it was during 1999 that the government's hESC/human cloning strategy began to take shape. Unlike the United States and Germany, where government bodies began to look into hESC after they had decided that, for the moment, all forms of 'cloning' were, in principle, unacceptable, things seem to have worked the other way around in the United Kingdom. Central policy actors saw no conflict between the HFE Act and human cloning for therapeutic purposes for research covered in the Act, mainly in the context of infertility treatment and reproductive medicine. By contrast, the combined issue of cloning and stem cell research was identified as associated with the vital rights of the British citizenry and the government defined as a defender of these vital rights. This position was officially acknowledged by the HGAC papers on the topic. Viewed from this perspective, hESC was a relatively unspectacular subtopic in the regulation of stem cell research technologies. But this view was in no way imposed from above but carefully presented and developed in a specific dramaturgy that involved the 'controlled' interaction with the public, in particular, the 'informed' publics created through consultation processes. These consultation processes addressed, on the one hand, a generalized 'British public' and, on the other, specific stakeholder publics with a special interest in topics such as cloning.

While there was considerable public concern about cloning, this technology was seen as a 'British technology', and further research in this area was not only a matter of regulatory decision-making but also a matter of defending important advances in medical research and of major industrial-strategic interest. From the late 1970s on, biotechnology had been defined as a major industrial-strategic goal for the United Kingdom. Following the Spinks Report in 1980 about the current state and future strategies for biotechnology in the UK, a plethora of biotechnology strategies and initiatives were designed (Gottweis, 1998, pp. 196–209). In 1996 the Department of Trade & Industry together with a number of other UK government ministries had launched a 'Crusade for Biotechnology' to maintain Britain's position in biotechnology into the twenty-first century. The idea of this 'crusade' was to maintain what was seen as Britain's lead in Europe and to strengthen its position as a global leader in biotechnology (*Biobusiness*, 1996, pp. 12–13).

The UK debate on stem cells and human cloning should be localized with this specific discourse on biotechnology, in which the British government was determined to 'defend the position of British biotechnology' in the world. Stem cell and cloning research came to be inseparably connected to a metapolitical goal of the Blair government.

But it must be pointed out that this is only one part of a larger story. The United States and the German governments were also committed to their national biotechnology industries, but the governments were much less certain about unequivocal support of human cloning and stem cell research. Within the group of core policymakers, the 'GMO scare' began to be articulated as an aspect of the stem cell issue in a very specific fashion. Not only was the stem cell issue seen and defined as a matter of major national interest, but it was also defined as an arena for demonstrating 'Britain's commitment' to the progress of science and its rejection of Luddites and irrationality.

In June 1999 the government announced the creation of an expert group chaired by the Chief Medical Officer, Professor Liam Donaldson, 'to advise on whether new areas of research could lead to a broader understanding of, and eventually to new treatments for a range of disorders where there is disease or damage to tissues or organs' (Department of Health (Great Britain), 2000b). The development of an expert group was met with some hostility in the scientific community, which feared a delay of advancement for British science and a possible brain drain to the United States. But at a press conference Donaldson dismissed such claims: 'We need to proceed carefully....I hardly think there'll be a brain drain in the next six months' (Masood, 1999, p. 4). The expert group was a small high-level body composed of 14 individuals, mostly scientists from reproductive biology and veterinary medicine, an ethicist, a lawyer and the government's chief scientific adviser, Sir Robert May.

With the launching of the group, led by Professor Donaldson, stem cell research and cloning finally had become a highly visible issue of governmental decision-making. It took the group much longer to produce and present the report than was initially indicated, the delay caused by several factors. For one thing, it seemed that the government wanted to have as much of a time distance between the BSE and the Genetically Modified Organism (GMO) crisis that had peaked in mid-1999. There was also an internal dispute within the governing party. The *Guardian* reported a row between the Department of Health and the science minister, Lord Sainsbury. The Department of Health was reluctant to publish the report and was concerned about possible critique from religious groups, but the science minister pressed to go ahead. Lord Sainsbury said in an interview that in his view, 'the potential medical benefits outweigh any other considerations one might have'. These remarks were interpreted as signalling the go-ahead for cloning. Liam Fox, the shadow health secretary, said, 'This is a huge issue of concern to church groups and religious groups, who are all expecting the very different ethical issues involved to be given maximum scrutiny.' Dr Fox, the *Guardian* reported, accused Lord Sainsbury of 'sweeping away all the complex ethical issues with complete contempt' and of having neglected the consultation that had been promised (Wintour, 2000). After almost a year of deliberations, the group issued its report in August 2000. It contained no surprises and, in essence, reiterated the policy content and direction of the 1998 HCAG consultation paper. In five chapters and two annexes, the report attempted to guide the reader through the main scientific, legal and ethical implications of stem cell research and human cloning, and it arrived at nine recommendations for future action. But the report was very clear that it did not attempt to take a fresh look at the moral and ethical issues of embryo research.

Quite clearly, the narrative of the report considered central ethical and moral issues of embryo research to be 'settled', and declared itself to be interested only in a set of 'new' questions. In short, in this staging of the hESC problem, the presence of any possible new moral or ethical question was simply and straightforwardly denied. No place existed for this type of already settled conflict to be restaged. In its recommendations, the Donaldson Report followed precisely the boundaries set up by the initial HGAC consultation paper: research using embryos should be extended from the strictly defined set of research areas dealing with infertility and reproduction to the general field of medical research, and cell nuclear replacement techniques should be used to develop treatments for mitochondrial diseases.

At the same time, research using embryos created by cell nuclear replacement should only be conducted if there were no other means to pursue the objectives of research. Furthermore, the report indicated a strict rejection of the mixing of adult cells with live eggs of any animal species and reproductive cloning. Moreover, throughout the report the earlier language of 'therapeutic cloning' had given way to the 'cell nuclear replacement' terminology. After the report was published, the government announced that members of the Parliament would have a free vote to decide on the proposed change in regulations (Radford, 2000).

With the Donaldson Report the British government had fully entered the politics of stem cells and human cloning. The debate on stem cell research moved from the terrain of 'generalized' and 'specific' publics to the field of Parliament as the site for deliberation and decision-making. The government stated that it accepted the report's recommendations in full and would bring forward legislation where necessary to implement them as soon as the parliamentary timetable allowed (Department of Health (Great Britain), 2000b, p. 1). The long duration of the production of the report, its presentation in the middle of the summer when Parliament was not in session, and the subsequent steps to introduce secondary legislation and finally pass the law were characterized by a series of unusual moves by the Parliament that indicated a great deal of insecurity and concern about the passing of the reforms. Just as was the case in the

1990 HFE Act legislation, until the final vote in the House of Commons in December 2000, it was uncertain what the outcome of this vote would be. After Parliament finally voted in support of the bill, it was far from clear how the House of Lords would proceed. However, the British case demonstrates an early, proactive political response to Dolly and the immediately arising questions of the boundaries of life, the focus of this response on hESC, the creation of an 'orderly' policy scenography with multiple actor roles, the concentration of possible regulatory action around the well-trusted institution of the HFEA, and the adoption of a deliberate style of interaction between various publics and the policymakers.

Dolly the demon

Neither 'Dolly' nor 'stem cells' have intrinsic meaning but take on meaning during the period that a relationship between these two terms is being established. To develop a policy strategy for hESC, one of the semantic links that was immediately created was to embryo research, as we saw in the case of the United Kingdom. In the United States, the evolving narrative on hESC was not only connected to the embryo research debates of earlier decades, but the mobilized discursive memory also explored a much larger field of questions dealing with the delineation of the boundaries of life. Unlike in the United Kingdom, no 'HFEA-like' regulatory agency was in place to deal with a topic like cloning, or to sort out differences between different types of cloning and hESC research. About a year before Gearhart and Thomson had come out with their results, in early February 1997, the birth of cloned sheep Dolly had caused a frenzy of political-legislative activities that anticipated regulatory positions taken by a number of participants during the stem cell debates. The various proposed cloning bills did not become laws, but they questioned, for the first time in decades, the clear boundary between publicly and privately supported research in the field of reproduction-related research (such as fetal, embryo and IVF-related research). The proposed laws applied on the federal level for all types of research and intersected with hESC in a number of ways. From 1998 on, in United States political discourse the topic of 'therapeutic' and reproductive cloning and hESC research collapsed into each other and created deep political division.

The question of cloning immediately became a high-level political topic, which was seen as requiring quick decision-making. On 24 February 1997, President Bill Clinton asked the National Bioethics Advisory Committee (NBAC) to discuss human cloning issues and to report with recommendations within 90 days (NBAC, 1997).

The president's call was the first step in a longer strategy to involve high-publicity ethics boards in governmental decision-making in the newly emerging politics of cloning and stem cells. The NBAC was established in 1995, its 15 members were appointed by the president, and its main function was to assess the appropriateness of departmental, agency or other governmental programmes, policies, assignments, missions, guidelines and regulations as they relate to bioethical issues arising from research on human biology and behaviour (Presidential Documents (US), 5 October 1995).

The extent to which cloning immediately became a highly charged topic was clear when, on 27 February 1997, the first anticloning bill was introduced in the US Senate by Senators Christopher Bond and John Ashcroft (US Senate, 1997). It stated that no federal funds may be used to finance cloning research. Many attempts to establish anticloning legislation were to follow (Forster and Ramsey, 2001, p. 204). In response to the action in Congress, on 4 March 1997, President Clinton issued a moratorium stating that 'no federal funds shall be allocated for cloning human beings'. The president also asked privately funded science to stop research. In its report from June 1997, the NBAC came to the conclusion that any attempt to clone human beings is unacceptably dangerous to the fetus and therefore morally unacceptable (NBAC, 1997). It recommended an extension of the moratorium. Following these recommendations, President Clinton introduced the Cloning Prohibition Act of 1997.

Following the February report that a sheep had been successfully cloned using a new technique, I requested my National Bioethics Advisory Commission to examine the ethical and legal implications of applying the same cloning technology to human beings. The Commission concluded that at this time 'it is morally unacceptable for anyone in the public or private sector, whether in a research or clinical setting, to attempt to create a child using somatic cell nuclear transfer 'cloning' and recommended that Federal legislation be enacted to prohibit such activities. I agree with the Commission's conclusion and am transmitting this legislative proposal to implement its recommendation.

(Presidential Documents (US), 1997)

Bills to ban cloning continued after these first attempts to draft anticloning legislation, but during the fall of 1997 these legislative initiatives seemed to lose momentum. On 8 January 1998, however,

Chicago scientist Richard Seed announced plans to clone a human baby, a development that led to renewed demands and initiatives in the legislative arena. Two of these initiatives deserve closer scrutiny, as they indicate the emergence of a 'new conflict about life', in which the fields of cloning and stem cell research were increasingly interconnected – an important nexus in the emerging policy narratives dealing with hESC research.

On one side of the political spectrum is the Human Cloning Prohibition Act of 1998 sponsored by Senators Bond, Bill Frist and Trent Lott (US Senate, 1998a), which would prohibit the use of SCNT for human cloning purposes. This measure would simultaneously forbid the creation of humans by means of cloning and research on cloned embryos. Furthermore, it suggested the creation of a National Bioethics Commission made up of theologians, philosophers, bioethicists, scientists and laypersons to study and recommend action on cloning. Penalties up to 10 years in prison and $250,000 fine were given to offenders.

On the other side of the spectrum is the 'Prohibition on Cloning of Human Beings Act of 1998' by Senators Dianne Feinstein and Ted Kennedy (US Senate, 1998b), which would ban the implantation of an embryo developed by the technology into a human uterus for the purpose of creating a child but would protect research on SCNT to clone molecules, cells and tissues. It would also pre-empt all state laws and contain a 'sunset' clause that would end the prohibition in 10 years. Under this law, the National Bioethics Advisory Commission would report to the President and Congress in 4.5 and 9.5 years on the scientific and ethical issues associated with cloning technology to recommend whether a federal moratorium should continue. Penalties for an intentional violation would result in a fine of $1 million or three times the gross gain or loss resulting from violations. The moratorium would expire after 10 years.

Neither bill ever became law, but they created links between crucial issue areas and thus mobilized broad constituencies 'for' or 'against' a new biological technology that promised to manipulate early phases of human life. The cloning controversy of early 1998 and the related attempts to structure the political space of research by establishing problem definitions and relationships between groups brought a vast number of actors into the political arena.

The early polarization of the hESC debate in the United States materialized in strong efforts by the supporters of hESC research to derail the fast-track anticloning legislation, a manoeuvre that turned into a major political goal for a broad variety of groups and associations and involved industry, medicine and science ranging from the Biotechnology Industry Organization (BIO) to patient groups and individual researchers, such as a group of 27 Nobel laureates. Two irreconcilable policy narratives seemed to emerge, with two very different scenographies describing the political action for the future. Against the background of Dolly and Dr Seed, supported by pro-life groups such as the NRLC (http://www.nrlc.org/news/1998/NRL2.98/doug.html) and backed by Senate Majority leader Trent Lott, the supporters of anticloning legislation had chosen the fast-track option with the implication to bypass committee hearings and debate. The debates in the Senate give a first impression of the strategies of argumentations and the narratives mobilized in the cloning debate but also later during the human stem cell discussions.

The anticloning legislation departed from the abortion arguments of the classical pro-life groups. The emphasis was not only on 'the destruction of life' but concerned something more abstract and generalized, 'life-related'. Opposition to cloning was increasingly defined as guided by the intention to 'protect the integrity of life'. Nevertheless opposition was strongly guided by the assumption that 'cloning' would be mainly a project of 'cloning babies for research', which was condemned by the pro-life forces. Support for cloning was increasingly equalled with the endorsement of research and medical progress. At the same time the discussion in the Senate was characterized by emotional discourse, in which the individual fates of senators, in particular, came to be key elements in the presentation of the cloning problematic.

During the discussion of the 'Human Cloning Prohibition' bill in Senate, the emerging positions in the struggle over life became clearer (US Senate, 1998c). Most strikingly, the debate in Senate was not a simple warm-up of the abortion debates under the heading of the cloning problematic. The critics of abortion attempted to discursively connect cloning not only to 'the destruction of life' but also to a more general 'philosophy of life'. On the other side of the political spectrum, the cloning topic was certainly not framed as a question of the rights of woman to choose. Instead, cloning research was powerfully connected with research, medical progress and the future of American science. In a letter to Senator Feinstein, for example, the American Cancer Society urged her to oppose the pending legislation by Bond and his colleagues.

With the Senate debates on cloning, many themes of the subsequent hESC debate were foreshadowed. Already at this time, potential applications of nuclear cell replacement techniques had become objects of legislative scrutiny. Without using the terminology, the 'Dolly debate' was about what later came to be called 'therapeutic cloning' in the United Kingdom, which supporters of research wanted to separate from 'reproductive cloning'. As we shall see in the following sections, with the advances by Thomson and Gearhart published in late 1998, further layers of complexity were added to the controversy. Nevertheless within 2 years, in the United States, the question of cloning had turned into a highly antagonistic, polarizing topic characterized by mutual mistrust and sometimes bizarre accusations for which any compromise seemed unlikely.

Stories stem cell research goes by: Between Catholicism and Buddhism

The previous sections have contrasted the unified proactive approach towards stem cell regulation in the United Kingdom with the defensive, reactive and fragmented US approach. Looking at these two countries, at first glance, it might seem that the US experience reflects the fragmented structure of the US political system, and the UK experience the centralized character of the UK system. The contrast between these two countries also seems to support the argument of the importance of radical religious groups in the political spectrum, which seems to be more prominent and influential in the United States than in Britain. But, as we have argued, such an interpretation would constitute a simplification. While these factors have played their role, the actual staging of the cloning controversy against the background of previous experiences to deal with such topics had played a crucial role in depicting and understanding the cloning problematic. In the following section we will further develop this argument by looking at Italy, Germany, Japan and South Korea. The relatively centralized character of the political systems of South Korea and Italy was not a guarantee of proactive regulation, nor was religion in Japan and Italy decisive in shaping regulation. Other factors played an important, additional role in shaping emerging strategies to 'deal with Dolly'.

Discursive memory also played a key role in the development of the German reaction to the development of cloning technologies. In the United States and the United Kingdom, the birth of Dolly had resulted in broad-ranging debates on cloning as a new technology with potential application in the human biomedical field, but such a framing was virtually absent in Germany. The mass media reported widely about Dolly, and the news of the first cloned sheep was met with shock and horror in large sectors of the German public, but its birth had not created any political-regulatory pressure for the simple reason that the existing strict German Embryo Protection Act explicitly outlawed cloning under strict penalty.

The German Embryo Protection Act is inseparably linked to the abortion question. In 1871, §218 of the German penalty law had declared 'intentional abortion' illegal with a sentence of up to 5 years in prison (Jütte, 1993, p. 22). Section 218 was slightly modified in 1927 by a court decision that in a case of balancing the life and health of the mother against that of the unborn, the life of the mother had a higher value than that of the fetus. With this decision a legal basis was created for medically indicated abortions (Jütte, 1993, p. 170). In 1976, after a long conflict between those in favour of following the liberal model of abortion law established in such countries as the United Kingdom and those rejecting a reform of §218, the parliament passed a law that permitted abortion in the first 12 weeks for reasons of medical necessity, sexual crimes or serious social or emotional distress. It required the approval of a second doctor, mandatory counselling and a three-day wait after counselling. With this law Germany had one of the strictest abortion laws in the world. In East Germany (then the German Democratic Republic), legislation was enacted in 1972 to permit abortion on request during the first 12 weeks of pregnancy. That law stipulated that 'in addition to the existing possibilities of contraception, a woman shall have the right to decide on the interruption of her pregnancy on her own responsibility, so as to be able to control the number, timing and spacing of births'. Negotiators working on the reunification of East and West Germany in 1990 were required to try to reconcile the differences between the two countries' laws on abortion. Finally, after a two-year legal debate on the issue, a new German abortion law was approved by the Bundestag (parliament). The 1992 law permitted first-trimester abortions on request after mandatory counselling and a three-day waiting period. However, before the new law could take effect, it was challenged in court by Chancellor Helmut Kohl and other conservative members of parliament, as well as by the State of Bavaria. In 1993 a confusing judgement by the Federal Constitutional Court found abortions were unlawful, since the constitution protects fetal life from the time of conception. Yet it also found that abortions during the first 3 months of pregnancy should not be punished if the woman first submits to counselling aimed at changing her mind. And, said the Court, such abortions are not covered under statutory health insurance and cannot be performed in state hospitals.

The German debate and controversy about fetal and embryo research must also be located within a highly complex discursive economy. Fetal research that was not targeted by the abortion opponents is relatively clearly separated from the discourse about embryo research. While the former is not regulated by any special legislation, governed by regulations of the Bundesaerztekammer (National Chamber of Doctors) and federally and privately supported (Schneider, 1995), embryo research is forbidden by the 1990 Embryo Protection Act, part of penal law and probably the strictest law regulating embryo research worldwide. Thus, neither privately nor federally funded research is possible in this area. In the United States, the main issue in the controversy was the equally symbolic and material question of federal support for embryo research – but the rights of the researchers to conduct embryo research were not touched. In Germany, the central issue was the duty of the state to protect 'unborn life'. German legal discourse emphasized the social function of the state over considerations of the freedom of research (Iliadou, 1999).

The story of the 1990 Embryo Protection Act began with legal challenges to the abortion law that had been reformed in 1973. In this context, the German Constitutional Court had ruled that German penal law did not regulate the first 14 days of human development from fertilization until implantation into the uterus (Betta, 1995, pp. 82–5; Iliadou, 1999, p. 36), nor could it be applied to embryos existing separated from the mother, as is the case in the process of IVF. From the early 1980s a debate evolved that focused on the need to find a legal solution to ensure the protection of 'unborn life' and at the same time be consistent with the legality of abortion. Furthermore, by the mid-1980s evidence had accumulated that research with embryos prior to implantation was going on in Germany (Rosenbladt, 1988, p. 213). The lack of any regulatory instruments was widely seen as highly problematic. In this situation, the Bundesaerztekammer issued guidelines that regulated research with 'early embryos'. According to the new guidelines, these 'early' embryos (defined as organisms from fertilization to implantation) were now 'officially' available for research purposes. In addition, the guidelines made provisions to create embryos for research purposes under certain special conditions (Iliadou, 1999, pp. 41–2). The new Bundesaerztekammer guidelines soon came under fierce attack.

During the second half of the 1980s, a controversy over embryo research erupted in which two strategies to regulate embryo research competed: one was strategy based on self-regulation by the medical community and the other a strategy based on formal law. The major research organizations such as the Max Planck Gesellschaft and the DFG

(Deutsche Forschungsgemeindschaft) were in support of self-regulation. On the other side was a broad coalition cutting across political positions from radical Greens and feminists to conservative groups that had come into existence and rallied behind the call for the state to protect embryos from abuse, instrumentalization and destruction (Betta, 1995, pp. 81–125). The outcome of this long and highly emotional debate was the passing of the Embryo Protection Act, a strict law that essentially outlawed most types of embryo research in Germany and, by implication, any private or public funding of such work. At the core of the act, which is part of the penal law, were a number of restrictions: prohibited is artificial fertilization that has any other goal than to result in a pregnancy, a measure precluding the creation of embryos for research purposes. Furthermore, the law prohibited any manipulation of embryos that served other purposes than their survival. The law defined an embryo as any fertilized egg cell beginning with nuclear fusion and, furthermore, any totipotent cell capable of division and development towards a human being (Iliadou, 1999). German regulations also encompassed privately funded embryo research and imposed a criminal penalty. In the absence of any strong rights-focused discourse, a clearly drawn boundary separating private from public research had never materialized in Germany. The Embryo Research Act put certain possible practices in the context of assisted reproduction under penalty. But many possible areas of regulation, such as licensing IVF clinics, were left to states and the German state doctors' chambers. Throughout the 1990s, the need for a federal law regulating medically assisted reproduction was broadly debated but did not lead to legislation. At the same time, federal support of medically assisted reproduction existed, as long as the work was not prohibited by the strict Embryo Protection Act (Deutscher Bundestag, 2002b, pp. 46–50). Thus, initially the Embryo Protection Act constituted something like a 'defense' against developments such as reproductive cloning, as it was per definition of the act outlawed in Germany. At least for the time being, there was no pressure on the policymakers to 'respond to Dolly' or to develop a strategy to deal with the possibility of reproductive cloning. As we shall see, this constellation was not favourable towards developing a proactive approach in stem cell regulation.

Italy[1]

In February 1997, the Italian news agency ANSA was the first to break the news of the birth of Dolly (cf. Gallese and Toldo, 1998; Satolli and Terragni, 1998; Neresini, 2000). Soon, journalists, bioethicists, scientists and politicians started to raise the question of whether the same technology that had given birth to Dolly the sheep could also be applied to human beings. Thus, the focus of attention quickly began to shift towards IVF-related issues and general boundaries of life questions (Neresini, 2000, p. 371).

During the 1980s and 1990s, Italy had turned into something like an experimental laboratory in the field of reproductive medicine. Italy's first 'test tube babies' were born in 1983 (Neresini, 2000, p. 14; Flamigni and Mori, 2005, p. 14). Approximately 50,000 others followed in the following decades. In addition, techniques such as donor insemination and sperm banking had started to spread in the late 1950s (Bonaccorso, 2004; Cirant, 2005, p. 134). In 1994 the Italian fertility expert Severino Antinori gained global fame when he helped a woman aged 62 to become a mother, and in 1995 a major uproar had been provoked by the birth of a girl in Rome who had been implanted in the womb of the embryo's aunt more than a year after the death of the embryo's 'biological' mother (Keates, 1995). These stories about messy genealogical relations, maverick doctors and births against all odds had created the impression of Italy being the 'Wild West of reproductive medicine'.

In the absence of primary legislation, Italian laboratories were governed by secondary legislation, by a series of court decisions, by the informal but nevertheless powerful norms and truths of Italian society, as well as by attempts of Italian physicians and scientists to make these norms binding through professional guidelines. But this 'dispersed' mode of governance was *not* the product of a decision or a shared consensus that would suffice. It was the effect of a series of failed attempts to find a common ground on how to reorder these tricky issues.

What had seemed before the birth of Louise Brown to be given by nature and without alternatives was now open to social negotiations and political reordering. The regulatory vacuum in the field of reproductive medicine led to the raising of more persistent questions. For years, Italians seemed to have preferred to live with these open questions rather than search for a common ground on how to reorder the field. Yet with the birth of Dolly the sheep, this reluctance became suddenly unacceptable.

After years of inactivity, Dolly led to the issuing of two emergency ordinances that banned *both* human and animal cloning (Ministero della Sanità, 1997b) and the commerce of human gametes and embryos (Ministero della Sanità, 1997a), and induced Italian parliamentarians to 'dare' to finally reach a common ground on how to rule these unruly issues. However, despite the agreement on the urgency of a law, it took the Italian Parliament an additional 7 years to agree on its content.

Thus, 'Dolly' also had significant impact on the emerging Italian strategy to deal with hESC research. In a very specific way, the absence of proactive governance in dealing with boundaries of life questions had led to a blended focus on a set of issues such as IVF, reproductive medicine, animal cloning and hESC research, with the overarching feeling of 'things being out of control'.

Japan and Korea

'I am not versed in the creeds of Buddhism. But when I carry out research, I always check whether they square with the sublime spirit of the Buddha. If I am not sure, I will be unable to continue my work', the Korean stem cell researcher Hwang Woo-Suk once said in an interview with the *Korean Times* in 2005 (Tae-gyu, 2005). Hwang's statement is typical of a genre of argument that constructs a harmonious relationship between Buddhist religion and stem cell research and attempts to explain more liberal tendencies towards stem cell research in some Asian countries through the influence of Buddhist religion. But, as we will argue, a closer look at two Asian countries, Japan and Korea, reveals that Buddhist religion is only of limited use in explaining the evolution of the regulatory constellation in these two countries.

In Japan, attitude behind shaping a regulatory approach towards hESC research and cloning was far from laissez-faire and instead was characterized by extensive consultation and deliberation eventually resulting in a restrictive and proactive course in cloning and hESC research. As in most countries, the issues of human cloning had generated considerable media attention in Japan, and Japanese regulators saw themselves suddenly confronted with what seemed a public much more susceptible towards issues of regulatory policy in the field of life sciences and biomedicine. In Japan, there was a particular relationship between the 'creation of life' topic of cloning and one of the most intensely debated topics in recent biomedicine, the question of organ transplantation.

The introduction of rules and regulations for organ transplants in most countries has been relatively unproblematic, despite the fact that organ transplants were made possible through developments in molecular medicine, such as the development of potent immunosuppressive drugs. In Japan, if there ever was a bioethics debate, it was not about genetic testing or any of the issues that cause concern in most other

countries, but about organ transplants and, more precisely, the definition of brain death. After repeated failures by the Japanese Medical
Association and its patron, the Ministry of Health and Welfare, to draft
widely accepted guidelines, a special committee was created in the late
1980s that included a number of prominent critical intellectuals. Over
the course of its deliberation, the council failed to find a compromise,
leading to the extraordinary release of a 'minority opinion', drafted by
three members of the committee, opposing the consensus view. The
deliberations of the committee received an unusual amount of press
coverage, and during this time a vigorous public debate by opponents
and supporters of the proposed brain death standards ensued. Since
the committee could not find a consensus, and faced with widespread
public opposition, officials at the health ministry dropped the issue,
deciding to wait for public opinion to calm down before launching a
campaign, which would be coordinated with prominent politicians at
the ruling Liberal Democratic Party (LDP), to introduce a bill to legalize
transplants of solid organs from brain-dead donors (Feldman, 2002).

Following temporary closure of the public debate on organ transplants with the successful accreditation by the Japanese parliament of
the Organ Transplant Act of 1997, a number of regulatory and bioethics
issues related to biomedical research suddenly surfaced – either because
they had been delayed because of the Organ Transplant debate (which
was the case with issues such as the research use of human cells and
tissues or genetic privacy) or else because of entirely new scientific
developments (as was the case with animal cloning and the successful culture of ESCs). Most important, a new culture of 'openness' and
increased concern with issues of 'public understanding' and even 'public accountability' had emerged during the second half of the 1990s,
following a number of high-profile accidents in Japan's nuclear energy
research programme that had serious consequences for the Science and
Technology Agency (STA) and may well have contributed to the decision to merge the agency with the Education Ministry, a plan that was
finalized as part of a large government reshuffling in 1998. Further,
issues of biotechnology regulation or IP rights suddenly emerged as a
topic for international politics and became debated at G7 summits and
meetings between heads of state.

Since, at the time, no legal or regulatory framework for reproductive
medicine or embryo research existed (in fact, in most cases, there were
not even guidelines), there was no obvious way to react to the issue of
mammalian cloning within existing regulations. Nevertheless the policy
reaction to Dolly was swift. On 18 April 1997, the Science Commission of

the Ministry of Education, Science and Culture established a Life Science Subcommittee within the Special Research Area Promotion Committee. The final report of the Life Science Subcommittee examined the situation of research on cloning in Japan and other countries and made suggestions for the establishment of a regulatory system. The suggestion was to impose a ban on human cloning. Equally, while the Education Ministry quickly instituted a temporary ban on human cloning experiments at universities or university hospitals (or undertaken with funding provided by the ministry), there was no ambition on the part of education ministry officials to go further. By contrast, its linkage with the Council for Science and Technology (CST) made STA into the most suitable agency to deal with this issue. On 15 June 1998, the Cloning Subcommittee of the Bioethics Committee of the CST published an interim report on 'fundamental principles concerning cloning', which, again defined the production of individuals through cloning as a violation of human dignity, while at the same time, opening up the door for human cloning techniques as such as long as they did not lead to the production of individuals. Finally, the Diet, the Japanese parliament, passed a 'Law on Human Cloning' on 30 November 2000, with a ban on human reproductive cloning at its centre.

Dolly also made a strong showing in the South Korean polity.[2] Unlike Japan, where the government managed to enact a law banning reproductive cloning quickly, in Korea 'dealing with Dolly' was far more protracted, and it took until 2005 until relevant legislation was in place. While much of what has been going on in Korean hESC research had been overshadowed by the Hwang scandal, which also was partially responsible for the delayed passing of regulations, three aspects of the Korean situation should not be overlooked: first, there was a considerable public movement and engagement to deal with the question of cloning in Korea; second, there were initial signs that the government was prepared to engage with those critical voices and; third, the government's failure and lack of will to act decisively on the regulatory level, and the resulting regulatory vacuum, were important elements in the political crisis building up in the wake of Hwang Woo-Suk's fabrications of research results.

Relatively soon after the news about Dolly was announced, some NGOs presented the government with a petition for a prohibitory decree to be enacted against human cloning (Kim, 2005). Green Korea United and the YMCA gathered together and expressed their concerns regarding sheep cloning that may inevitably lead to human cloning. They urged the government to establish the National Bioethics Committee. Furthermore, The Catholic Bishop's Conference of Korea particularly presented 'a petition to enact the Human Cloning Experiments Prohibition Law' to the National Assembly, urging the prohibition of any tests or experiments in relation to cloning. The National Assembly also started legislative manoeuvres to ban human cloning, and in 1997 and 1998, members of the National Assembly proposed a revised bill, the Biotechnology Promotion Law. The characteristics of the two bills were to merely forbid human cloning, reflecting societal concern rather than to regulate research efforts at all. However, this legal exercise was also neglected without even a careful deliberation. Unlike the National Assembly, society could not remain placid. In 1998 Kyung Hee University Medical Center announced that embryonic cloning had been successfully executed, elevating the level of anxiety that human cloning would be only a matter of time.

In September 1999 an important event took place with respect to the cloning debate. Hosted by The Korean National Commission for UNESCO, a consensus conference on cloning took place engaging with the general public. A panel of citizens participating in the Consensus Conference concluded that human embryonic cloning as well as reproductive cloning should be banned (KNCU, 1999). A social consensus about hESC research was sought through the Citizen Panel Report, and it was proved that citizens in general were capable of making a proper assessment of abstruse issues in the fields of science and technology. Hwang participated on an expert panel. He stressed the need for human embryonic cloning. He even mentioned that he had already started cloning research using human cells.

What followed after the consensus conference was the slow process of drafting a bill that would eventually (in 2005) regulate hESC research and cloning in South Korea. A key role in this process was the civic group Center for Democracy in Science and Technology (CDST), which has consistently organized debating forums about biotechnology since 2000, with an intention to publicize the relevant issues to achieve social consensus. Among the debating forums that have been held – 'Necessity for Bioethics and Safety Law Workshop', 'Human Embryonic Cloning Forum for 14 Days', 'Protection of Genetic Information Forum' and 'Gene Therapy Forum' – were intended to scrutinize the current status and issues of each field. In particular, in October 2000, CDST presented the National Assembly a petition for a Biotechnology Safety Ethics bill. Having played a crucial role in the process of enacting a bioethics law, the backbone and basic contents of this petition included a ban on human cloning and unethical research, a restriction on gene therapy, a prohibition on genetic discrimination and the establishment of the

National Bioethics Committee. While the government focused on the potentially huge economic impact of cloning for the biotechnology industry, civic groups suggested alternatives and urged the National Assembly to enact the related laws.

Thus, the picture emerging from how South Korea dealt with the challenge of Dolly is ambiguous. On the one hand, a line of action developed that seemed to point towards a transparent and deliberative style of policymaking in this field. On the other, there seemed to be a clear lack of determination to implement this direction of governance. As we will show in the next chapter, the consequences of this non-decision-making are evident in the context of Hwang-gate, and the evolving political scandal in stem cell research.

Conclusions

We have argued in this chapter that the emerging regulatory regime in stem cell research was closely related to the dislocatory impact of the 'event' of the birth of the first cloned sheep, Dolly, and the strikingly different ways in which the phenomenon Dolly came to be interpreted, narrativized in different countries and put into the larger policy context, and how different policy dramaturgies emerged. The boundaries of life moved into the centre of policymaking, and the central issue was to which extent animal cloning triggered a larger social debate on the politics and culture of life or resulted in the creation of focused regulation. Different stories on cloning gave rise to different stem cell dramas, played out by different groups of actors.

Our comparison revealed that neither arguments of path dependency nor the autonomous operation of any cultural patterns have strong, exclusive explanatory value for understanding regulatory variations. There is, for example, no question that discursive memory and previous regulatory history, such as in Germany or in the United Kingdom, can favour certain regulatory pathways over others. There is no question that religion was a major factor in shaping regulation in Italy or Japan. But we are dealing here with highly complicated mechanisms of decision-making, which are not easily reduced to specific, single factors. In all countries under consideration, the major industry policy actors had long ago defined biotechnology as a key industrial resource to be supported by all means through government. While this argument could become central in South Korea, it remained peripheral in Germany, and was only occasionally brought into play in the United Kingdom. Whereas in Korea the stem cell drama from the very beginning was about the country's biotechnology industry and economic greatness, in Germany it centred on human dignity, and in the United Kingdom it focused on a pragmatic decision for a good cause. No evidence exists that religion could play an independent role in shaping decision-making in any particular direction in any country. While the Buddhist religion definitely has different concepts of what constitutes human life than Christian religions, we could not identify any regulatory patterns where this aspect played a central role. For example, for years, Italy, the parade Catholic country, espoused the most liberal regulation in reproductive medicine and stem cell research in the world. From one year to the next, this picture changed dramatically. Also, it would be misleading to believe that Buddhist religion unilaterally embraces abortion and has an undifferentiated view on questions of life, abortion and embryo research, or that Buddhism necessarily has a more generous interpretation of stem cell research than Christianity. The cautious Japanese approach is proof of that.

These insights point at two, partially contradictory, conclusions for the regulation of regenerative medicine beyond the hESC conflict, such as questions of transplantation or human gene modification. First, they show that the mobilization of cultural patterns and previous regulatory history play an important role in the shaping of policy scenographies and resulting regulatory regimes. Second, they point to the fact that the particular mobilization of these regulatory paths and of cultural patterns, their narrativization and dramaturgy do matter, and, in fact, play a key role in the creation of regulatory variance. Thus, dealing with cloning displayed rather different features in our comparative perspective: in the United Kingdom and Japan, the focus of regulatory efforts was on a narrativization of cloning as an important topic that, nevertheless, could be dealt with by utilizing well-established regulatory mechanisms. Furthermore, in most countries, existing or still-to-be-created bioethical institutions began to be positioned as important sites for deliberating hESC and cloning research. In general, cloning was constructed as a 'normal' topic that required diligence, a certain amount of openness and the creation of trust. In the United States, Germany and Italy, cloning quickly became a 'big topic', a state affair that polarized the respective policy communities. In Korea, until the late 1990s, an effort of broad public deliberation was made to deal with the cloning topic. But neither the United States, Germany and Italy nor Korea developed early policies regulating cloning and, later, stem cell research. The generation of trust for governmental regulatory policy was a highly complicated project. Our first insight points at the limits of any global regulation in stem cell research, and the next chapter will show how the mentioned differences in narrative mobilization led to difference in regulatory structures. However, the second insight, that dramaturgy and the policy 'mis-en-scene' do matter, relativizes the independent impact of existing political-structural or cultural patterns, which raises the scenario of the actual possibility for transnational regulation. Consequentially, in Chapters 5 and 6, we will discuss in detail the role of bioethics as a new language and practice when dealing with regulatory challenges stemming from regenerative medicine.

4
From Dolly to Therapies? Stem Cell Regulations in the Making I – The United Kingdom and the United States

Introduction

Although announcements of the birth of a cloned sheep, a baby born to a 62-year-old woman, plans by a researcher to try to clone a human baby, as well as the prospect of reproductive cloning, had all sent considerable shock waves throughout the Western world, none of the countries discussed in this book had clear majorities that supported a total ban on therapeutic cloning, or on hESC research in general. By the late 1990s it was far from certain that, for example, the United Kingdom would embrace hESC research, or that the United States would, at least on the federal level, develop a restrictive approach. As we will see in this chapter, much depended on careful narrative maneuvering, framing, interpreting and the construction of stories and scenographies in which the growing conflicts between a variety of groups and individuals were enacted. After 1998, regulatory frameworks began to take shape worldwide, with significant differences displayed across nations. In the last chapter we have shown how these differences in regulation originated to a considerable degree in contrasting strategies to deal with the dislocatory event of the first cloned sheep Dolly, the adoption of different tactics to respond to the consequent issues about the boundaries of life and the buildup of strikingly different policymaking dramas. We will continue our discussion by looking closely at the shaping of hESC regulations in the United Kingdom, the United States, Germany, Italy, Korea and Japan.

United Kingdom

As discussed in the previous chapter, in 1999, the government had commissioned the Donaldson Report, which focused on a set of 'new' questions and declared many of the ethical and moral questions surrounding embryo research 'settled'. The findings echoed many of the recommendations of the HGAC consultation paper, which included extending the use of embryos in areas other than infertility and reproduction and using cell nuclear replacement techniques to develop treatment for mitochondrial diseases.

The British government's approach did not reflect an already existing social consensus but instead was a proactive attempt to structure a highly specific policy scenography and course of action. It is crucial to see that it was hardly the presence of the HFEA that unquestionably led the way into a new legislation, but a careful and extensive process of narrative maneuvering (Human Fertilisation and Embryology Act, 1990). Critics of the proposed legislation existed and were far from negligible. They included a number of smaller pro-life groups whose narrative fiercely rejected the proposed legislation, the 1990 HFE Act and the 1976 abortion legislation. These included not only LIFE and SPUC, groups that had already played an important role in the debate of the HFEA legislation, but also newer groups, such as Comment on Reproductive Ethics (CORE, founded in 1994) and the ProLife Alliance (founded in 1997). Other critical organizations included the NGO Human Genetics Alert (London) and the Catholic Church (Kamal and Hinsliff, 2000). More salient in a political sense, critics also comprised representatives from all political parties in Parliament, who either sided with a pro-life agenda or had other reasons to reject the proposed legislation. The Catholic Church, although repudiating stem cell research and cloning, remained relatively silent and inactive during the legislation's debate. The ProLife Alliance, however, and LIFE and SPUC, to a lesser extent, became the most articulated and active voices on the pro-life side. These three groups provided a countless stream of press releases, statements and reports, and created a strong presence on the Internet through well-organized websites.[1] The most recent of these organizations, the ProLife Alliance, in fact, declared itself a political party and fielded candidates in the 1997 elections. It announced its central goals as the repeal of the 1976 Abortion Act and the outlawing of all abortion; the repeal of the 1990 HFE Act; the prohibition of cloning, embryo experimentation and destruction, and the laboratory creation of embryos and; the banning of voluntary and involuntary euthanasia (ProLife Alliance, 2001, p. 3). These pro-life organizations and the Catholic Church believed that all forms of embryo research and cloning were fundamentally wrong.

In the critics' story, a conflict between the pro-life groups and 'The Government' existed and suggested a relatively clear structure of the world, which was divided among those who rejected abortion, stem cell research, and cloning – and those who supported it, a division that separated 'Christians' from 'Non-Christians'. In the words of one of the Pro-life leaders:

> The United Kingdom is a profoundly Post-Christian culture. There is definitely no longer any concept at all about Christian heritage, of issues of right and wrong, and in many respects, it is my belief, science has usurped the role of religion ... and scientists are our Gods, whatever they say is the truth.... The whole debate polarized into science versus religion and this was very much borne out of this: there is science, and there is religion.... We never have come to any conclusion what the human embryo is, but the arguments were looked in 1990 subsequent to the time the Warnock committee rather brushed aside laws and regulations, and we were just left alone with that.... God has gone from this country.... It is a country without a soul, totally hedonistic, self-centered culture.
>
> (Interview, London, 4 July 2002; see also Warnock Report, 1985)

But these views, perceptions and 'divisions of the world' entered the political debate only in a rather limited way. This political debate was dominated by an alternative semantic construction that interpreted the 'stem cell/cloning problem' as one in which pragmatism conflicted with dogmatism. The HGAC consultation paper and the Donaldson Report, the increasingly dominant interpretation of 'the problem' suggested that the well-respected HFEA system, and a few, but necessary, modifications were all that was needed to extend the scope of permissible research from fertility and reproductive medicine topics to other areas of medical research, such as Parkinson's and Alzheimer's disease. What could be more natural and obvious? In fact, 'there was no problem' – all potential problems, such as the status of embryos, had been discussed and settled many years ago. This 'pragmatic view', shared by a network of important and respected groups, institutions, and individuals in Britain, was deemed to be misinterpreted and perhaps misunderstood by a small minority who preferred to stick to old dogmas and views, rather than accommodate to the new situation. In this narrative, the 'small minority' had respectable values and should not be ignored, but these critics were simply misled and thus had misinterpreted scientific and political realities.

In the evolving staging of the regulatory process, the critics – though potentially comprising large segments of the population such as the

Catholics– were constructed as a small group. As one member of the Wellcome Trust put it: 'We were aware that anything in this country that involves any human embryo always creates fairly strong opposition of fairly small groups, in particular, the pro-life alliance.... If I am honest, that seems to be the only focus of opposition, is that small group.... It is not clear how many people they represent' (Interview, 4 July 2002). However, government caution in preparing and passing the new HFE Act legislation indicated that this small group's concerns were taken very seriously. The government did not want either the 'BSE scare' or the 'GMO experience' to be repeated in the case of stem cell research. The discursive memory of past mistakes fully entered into the developed policy strategy.

The supporters of stem cell research comprised the government; representatives from all political parties in Parliament; medical, scientific and industrial organizations; patient groups; and a wide range of academic bodies. This broad alliance included, among others, the Royal Society, the British Medical Association, the Medical Research Council, the Biotechnology and Biological Sciences Research Council, the Human Fertilization and Embryology Authority, the BioIndustry Association, the Association of Medical Research Charities, the Alzheimer's Society, the British Liver Trust, the Parkinson's Disease Society, the Juvenile Diabetes Foundation and the British Heart Foundation. Also the Nuffield Council on Bioethics, Britain's most important voice in bioethics, came out with a discussion paper supportive of hESC research (Nuffield Council, 2000). This powerful set of actors nonetheless could hardly be sure of its victory in this highly sensitive conflict.

In its evidence to the Chief Medical Officer's Expert Group, the Royal Society recommended that a working party should investigate the feasibility of establishing frozen banks of various categories of stem cells that have been both tissue-typed and screened comprehensively for pathogenic viruses. In response, the Expert Group recommended that the research councils should be encouraged to establish a programme for stem cell research and should consider the feasibility of establishing collections of stem cells for research use.

'We are glad that the Expert Group has endorsed our recommendation,' said Professor Bateson from the MRC in a press release. 'Much more research is needed on all types of stem cells if patients are to benefit from the new therapies that will be made possible by such work' (The Royal Society, 2000a). This statement indicated a considerable shift in the Royal Society's position towards hESC research and particularly in cell nuclear replacement. In February 2000 a working group chaired by Richard Garder produced a paper on 'therapeutic cloning', intended to contribute to the deliberations of the Donaldson Committee. In its submission, the Royal Society discussed the possible benefits of therapeutic cloning as a potential basic research tool but concluded in a highly sceptical manner:

> The techniques of therapeutic cloning [are] likely to remain inefficient for the foreseeable future, and ... raise serious issues about safety, particularly regarding the normality of donor nuclei. If this approach for replacing damaged tissues does work, the cost will be considerable. This may mean that such therapy will only help those individuals who are able to afford an expensive treatment and the majority of patients will be excluded. Therefore, the early applications of these techniques are likely to be offered by private clinics.
>
> (The Royal Society, 2000b)

Such considerations were absent from subsequent statements of the Royal Society, including its briefing prepared for members of Parliament in November 2000 as the decisive vote about the proposed legislation came closer (The Royal Society, 2000a). The brief emphasized the crucial importance of stem cell and cloning research for the future of medical research and the position for British science in the world. The Royal Society's position shift indicated that the UK research establishment had decided to attribute not only great substance to the 'battle over stem cell and cloning research' but also great symbolic importance. As an MRC representative put it:

> This was one issue we wanted to make a stand on for rational approaches to science, rather than this anti-science lobby we often get in the UK. For example, if you take GM crops, scientists assumed ten years ago that the case was so overwhelmingly in favor of using GM crops they did not bother to do anything, and then Monsanto came along and actually made a few inappropriate remarks and comments. The whole GM cause got lost, and I think everyone is determined with things like stem cells – that is one thing where we have learnt ... that did not get involved because they thought it is so obvious everyone looks at it, but of course, that was naive, and so people were very clear let's get out there and let's make a case, let's get to the press, tell the press what is going on, and lets get the arguments out there, let's get people like Christopher Reeve, Superman, telling the general public what the potential benefits are rather than the

scare stories, and let's make sure that scientists get out there and tell the truth about reproductive cloning. Nobody in their right minds believes that reproductive cloning is something that one could contemplate or could even be done scientifically.... Stem cell science finally had turned into a symbolic battlefield where the future of British science was to be negotiated.

(Interview, London, 3 July 2002)

From August to December 2000, from the publication of the Donaldson Report until the decisive vote on the House of Commons, the medical, scientific and industrial organizations in support of stem cell and cloning research, along with medical charities and patient groups, organized an unprecedented, well-coordinated campaign in support of research, organized by the Wellcome Trust, a medical charity (Interview, London, 5 July 2002).[2] In addition to the Trust, the MRC, the Department of Health and other organizations organized briefings and meetings with MPs to communicate the focus of the evolving policy narrative: 'There is no problem. Pragmatism must win over dogma.'

Pragmatism versus dogma

In the United Kingdom, just as in many other countries, hESC regulation was seen as requiring legislative action, and thus the debate moved at some point from the dispersed local arena of briefings, consultations and media releases to the legislative stage.

A number of themes emerged during this debate and the subsequent parliamentary debates in December 2000. These themes reflected the strategic priorities of the supporters of the impending legislation, but also, to some extent, of the critics. A key strategic decision of the proposed legislation's supporters, beginning with the HGAC consultation paper and the Donaldson Report, was to avoid 'reopening' the debate that had taken place in Britain around the publication of the Warnock Committee report leading up to the 1990 HFEA regulations. In the dominant policy narrative, questions of 'what is life' or about the general acceptability of embryo research should not become the centre of concern in the parliamentary debate. And, indeed, the House of Commons debates did not focus on this topic. These 'general questions' had become non-issues, which certainly did not disadvantage the supporters of research. As one of our interviewees put it: 'The debate did not go back and reopen as to whether or not human embryos should be used in research or not, and this is why I think when it came to debate of stem cells and their generation from human embryos, things moved easier in the UK, much easier than they did in Germany or in the States, because to a certain extent we resolved the use of human embryos in research argument ten years ago' (Interview, Terry, 4 July 2002). The MPs and scientific and other organizations that supported stem cell research were clearly saying, 'There is, in fact, no problem. New legislation is needed to help those with serious diseases.' This focus had become so dominant in the months and weeks leading up to the parliamentary debate that critics of stem cell research had resorted to a defensive strategy. They questioned the claim that the proposed reform was really no more than incremental change; they complained about the lack of time given to Parliament to come to a decision; and they pointed to the potentially disastrous consequences of a positive vote by Parliament. This strategy was enacted through public statements, press releases and other forms of dissemination; then through MPs in Parliament; and finally in the form of court action intended to find judicial support for the core accusations of the stem cell and cloning research critics.

One clear line of argument in support of the new legislation was that people with serious diseases must be helped. Announced changes in the HFEA were presented on 27 November 2000. Revised draft regulations were issued on 12 December 2000, in particular, involving a rewording and a limitation of the purposes of the act to 'serious' disease. This was a response to concerns articulated in the previous parliamentary debate that embryo research might be permitted also for minor ailments or trivial complaints. The new regulations extended the scope from permitting research in embryos and IVF not only for infertility treatment and the study of the development and treatment of embryos but also for a sixth area of increasing understanding about human diseases and disorders and their cell-based treatments.

On 15 and 19 December, the decisive debates in Parliament took place. Parliamentary Undersecretary of State for Health Yvette Cooper introduced the debate: 'The purpose of the regulations is to promote stem cell research, which has immense potential to relieve the suffering of many people in this country. It is for that reason, and because of the impact that the research could have on hundreds of thousands of people, that the Government supports the regulation' (Cooper, House of Commons Debate, 2000).

Many speakers supportive of the new legislation emphasized the enormous medical potential of stem cell and cloning research. In the decisive vote, the House of Commons gave a clear vote: 366 for the proposed legislation, 174 against it.

After the vote, the proposed legislation moved to the House of Lords. One day after the vote in the House of Commons, the government announced its intention to bring the statutory instrument to the House of Lords one day after the beginning of the next parliamentary session. When protest rose against such haste, the debate was postponed to 22 January. Pressure continued against the planned legislation. The Archbishop of Canterbury joined forces with the Roman Catholic archbishops of Glasgow and Westminster, the Chief Rabbi and the president of the Muslim College to protest changes to the law that would allow testing on stem cells derived from the cloning of human embryos. In an open letter that was sent to all peers, the religious leaders said the proposals deserved to be examined in more detail than was allowed in a brief parliamentary debate (*Guardian*, 2001a).

Again, the outcome of the vote in the House was considered ambiguous. On 22 January the House discussed the Human Fertilization and Embryology Regulations 2000. Before the discussion Lord Alton had tabled an amendment that a Select Committee should be appointed to study the proposed regulations. This proposal would have delayed a final decision by the House for up to a year. In response, Lord Walton of Detchant tabled a different amendment that the bill be approved but that a Select Committee would then scrutinize the new regulations.

However, the supporters of the Walton amendment pointed out the need to act quickly. In the final vote, Lord Alton's amendment was rejected (212 votes to 92). The alternative amendment of Lord Walton, which proposed creating a 'retrospective' Select Committee was passed without division. The new regulations came into effect on 31 January 2001. Although legislation was indeed rushed, the chosen strategy reflected the government's determination to have appropriate regulations in place as quickly as possible and to move ahead with research.

But this was not the end of the story. After allegations in Parliament that legislation had been rushed through, the ProLife Alliance applied for judicial review, which was rejected. Under the chairmanship of the Bishop of Oxford, Richard Harris, the House of Lords began its exhaustive investigations into stem cell and cloning research. Finally, the House of Lords' Committee came to conclusions that fully endorsed the 2001 HFEA and the 2001 Reproductive Cloning decision. The positions and oral and written evidences not surprisingly repeated the views and arguments advanced by the various groups leading up to the 2001 legislation. On 27 February 2001, The House of Lords Select Committee officially announced its findings. At a press conference, Richard Harris said, 'Stem cell research offers real and great hope for the future for a range of common diseases.... It is very important to keep all avenues of research open and we recommend that for the moment this fundamental research is necessary.... The Human Fertilisation and Embryology Authority ... is the best regulatory authority in the world' (*Guardian*, 2001b). Two days after the House of Lords decision, it was announced that HFE Act had granted the first licences to carry out stem cell research under the new regulations to the Edinburgh's Centre for Genome Research and a group of researchers at King's College, London (*Guardian*, 2002). While this course of action certainly was built upon earlier embryo legislation and the presence and ethos of the HFE Act, the careful staging of the hESC debate reflects a proactive, coherent governance approach whose implementation was the result of a broad political and semantic mobilization.

United States

In the United States, the birth of Dolly had created a strongly polarized division within the political landscape. The political actors were required to operate within a confusing regulatory context in which the boundaries between abortion, fetal and embryo research and cloning were rather blurred, and the separation between privately and publicly supported research in the field of reproductive medicine and embryo research was unusually stark. Contrary to much of what had been written about the US hESC strategy, we argue that the stem cell governance strategy evolving between the Clinton and the Bush administrations reflected an attempt to develop a proactive strategy. As we will show, this strategy reflects a long, gradual process of separating embryo and fetal research from abortion politics and associating these fields with the idea of medical progress, a process very similar to the developments in the United Kingdom – but with a different political outcome.

Redefining research on embryos

This shaping of the US policy scenography was influenced considerably by the rights-focused discourse about the regulation and public funding of non-therapeutic abortions. The central argument in this context was that the absence of public consensus regarding the moral status of the 'preembryo' excluded both the development of regulations constraining research and the use of public funds to support such research. In either case, the liberty or resources of some individuals would inappropriately be constrained or co-opted to pursue ends that they would explicitly eschew (Khushf, 1997, pp. 497–8). Although federal funds for many kinds of fetal and embryo research were not available, privately sponsored research continued, either, as in the case of embryo research, without the existence of any viable regulatory system, or, as in the case of fetal research, within a complex system of federal and state laws. Thus, the rights-focused political discourse had played a critical role in justifying and explaining the creation of a boundary between private and public in the field of fetal and embryo research. It articulated a specific form of biopolitics in which definitions and potential uses of life were no longer the exclusive prerogative of the state and legislation, but partially left to the forces of a market composed of entrepreneurs, scientists, clinics and medical doctors. As we will see, it was precisely this policy of state non-interference in embryo research that gradually began to change under the Clinton administration, preparing the way for federal support of hESC research in the late 1990s. Cloning policies and controversies continually affected this development, but it was part of the Clinton's and Bush's administration strategy to try to separate questions of SCNT from hESC research.

In 1993, the NIH formed the Human Embryo Research Panel (HERP), which set forth guidelines delineating appropriate and inappropriate areas of embryo research (Tauer, 1997). In one part of the report, the panel made the controversial decision that researchers should be allowed to create embryos for certain research purposes. This position was rejected by President Clinton, who stated that federal funds should not be used to create embryos for research purposes. During the appropriations process for the budget of the Department of Health and Human Services (DHHS), Congress stipulated that an activity involving the creation, destruction or exposure to risk of injury or death to human embryos for research purposes could not be supported with federal funds (Kinner, 2000a, p. G4). In a 1995 rider to the appropriations bill co-authored by Republican Jay Dickey (later called the Dickey Amendment), any research was prohibited posing a risk to an organism derived by fertilization. Despite this de facto ban on federally funded embryo research, the private sector continued to be free to make up its own rules and to conduct research. Thus, until the early 1990s the basic US approach to fetal and embryo research was not to impose regulatory restrictions but also not to spend public money on it.

Thus, in the United States, with Congress hesitant to fund research involving human embryos, most embryo and IVF research had been driven into the hands of private corporations. By the end of 1999, all research on totipotent and pluripotent stem cells in the United States had been conducted in the private sector, with the key research groups at the University of Wisconsin and at Johns Hopkins University being supported by Geron Corporation. The research that aspired to create human embryos with cloning technology was financed and carried out by Advanced Cell Technology, a biotechnology company based in Massachusetts. This development of embryo research, mainly financed by private companies, gave rise to a controversy focused on if and under which conditions ESC research should be supported by the government (Annas et al., 1999; Frankel, 2000). However, the main participants in the political debate failed to clearly point out that no federal regulatory system existed to deal with embryo research and IVF. Likewise, the boundary that separated private and public research in this area was not questioned by the main actors in the controversy.

The privately supported research of Thomson and Gearhart became both a symbol for the substantial medical potential of ESC research and a demonstration of what many in the medical-scientific community saw as the shortcomings of embryo research in the public realm. Whereas Gearhart's work could have been supported by government money, Thomson's research involved abstracting cells from a blastocyst and then destroying the embryo. Thus, it would have been excluded from federal support.

Only weeks after the announcement of the successes of Thomson and Gearhart, the stage was set for the forthcoming controversy. Although Harold Varmus, the director of NIH, said that ESC research 'is of sufficient magnitude and importance that the federal government should be playing an active role in supporting it', Rep. Jay Dickey defended the ban on embryo research, asserting that 'the ban serves a very good purpose in our society because it honors the sanctity of life' (Butler, 1998, p. 104). At the NIH, the view increasingly predominated that it was not a question of whether the government should support hESC research but how to justify and explain the necessity of it (Interview, NIH, Bethesda, MD, 30 August 2000). A new, proactive strategy in hESC regulation seemed to take shape. In this constellation of conflict, the critical challenge for all of the central actors was to craft a policy narrative that would eventually offer the dominant interpretation of the nature and relevance of hESC research, create a new policy dramaturgy and mobilize a powerful coalition promoting and justifying government support in this field.

In light of the presidential and legislative bans, the NIH requested a legal opinion from the General Counsel of U.S. Department of Health and Human Services (HHS) on whether federal funds could be used to support research on human stem cells derived from embryos or fetal tissue. HHS's General Counsel, Harriet Rabb, concluded that the current law prohibiting the use of HHS-appropriated funds for human embryo research would not apply to research using stem cells 'because such cells are not a human embryo within the statutory definition'. General Counsel Rabb determined that the statutory ban on human embryo research defines embryo as an 'organism' that when implanted in the uterus is capable of becoming a human being. The opinion stated that pluripotent stem cells are not and cannot develop into an organism, as defined in the statute. HHS concluded that the NIH could fund research that uses stem cells derived from the embryo by private funds, but because of the language in the rider the NIH could not fund research that derived stem cells from embryos using federal funds (Rabb, 1999).

An important step in the creation of a policy narrative explaining the importance of ESC research was President Clinton's request in November 1998 to the NBAC to prepare a review of the medical and ethical issues associated with human stem cell research. After the 1997 Cloning Prohibition Act (Presidential Documents (US), 1997), this was a crucial move not only to separate cloning issues from hESC issues but also to develop a proactive government strategy. In January 1999, the director of NIH imposed a moratorium on research using human pluripotent stem cells derived from human embryos and fetal tissue. In September 1999, the NBAC delivered its two-volume report to President Clinton in which the Commission offered a 'political philosophy' of stem cell research (National Bioethics Advisory Commission, 2000). The Commission considered it as its task to produce a statement in which a majority of the Americans should see their concerns and considerations reflected (Interview, Hastings Center, Garrison, NY, 1 May 2000). After a thorough discussion of the ethical, scientific, legal and social implications of stem cell research, the NBAC strongly endorsed the funding of hESC research. It advocated that the derivation and use of human cells from cadveric fetal tissue should continue to be eligible for federal funding, but also the derivation and use of hESCs from embryos left over after infertility treatments. At the same time, the Commission recommended that federal agencies should not fund research involving hESCs derived from embryos made solely for research purposes using IVF. Furthermore, it suggested a tight system of national regulation, with the DHHS establishing a 'National Stem Cell Oversight and Review Panel' (pp. 3–7).

Already, during the deliberations of the NBAC, the question arose of whether federal funds could be used for research on stem cells using 'surplus' embryos from IVF treatments. The legal opinion of General Counsel Harriet Rabb of DHHS on the acceptability under current federal law to support research with stem cells was another important step in the shaping of the evolving policy narrative. Rabb had concluded that pluripotent cells are not organisms and they do not have the capacity to develop into an organism; on the contrary, they have the potential to evolve into different types of cells, such as blood cells or insulin-producing cells. Furthermore, Rabb introduced a distinction between the destructive removal of ESCs from embryos and research that occurred with those cells after their removal. In Rabb's view, federal research support could be provided as long as this research did not involve the removal of ESCs (Robertson, 1999, p. 112). Hence, a key figure in the legal-political system had made the decision that the kind of things Thomson and Gearhart were working with were not embryos or something related but, rather, 'simple cells', which therefore had no 'life qualities' (Rabb, 1999). Rabb's semantic creation of two types of stem cell research and her political framing of the status of the embryo would soon become a central axiom in the gradually developing stem cell policy narrative.

The director of NIH, the president of the United States, the NBAC and Representative Dickey were hardly alone in the emerging discursive space of ESC research. A growing number of actors began to influence the policy-problem definition process. The efforts to develop a government position on federal support for ESC research culminated in the NIH draft guidelines for research using human pluripotent stem cells. It was released in 2 December 1999 and received approximately 50,000 comments from members of Congress, patient advocacy groups, scientific societies, religious organizations and private citizens (National Institutes of Health, 1999b). The final guidelines were published on 25 August 2000 (National Institutes of Health, 2000b).

Central to the new guidelines was the decision to extend the scope of government support for ESC research and to create a framework for regulatory oversight. In the past the law permitted the use of federal funds for the derivation of ESCs from aborted foetuses, but the situation was less clear with respect to the crucial issue of spare embryos donated by couples undergoing IVF. The 25 August 2000 guidelines clarified this ambiguity, permitting the use of cells derived from IVF spare embryos for research purposes. Most important, the guidelines constituted a forceful, unambiguous statement of support for ESC research.

The Clinton administration's emerging policy narrative also featured a number of references to more general US political metanarratives.

With American health and American leadership, two core elements of contemporary US political identity were mobilized and connected to ESC policy. Various statements in government reports and guidelines concerning stem cells left no doubt that the continuation of stem cell research in the United States would be critical to help Americans fight dreadful diseases for which no cure is currently available (National Bioethics Advisory Commission, 2000; National Institutes of Health, 1999a, updated 2001). In other words, the health and the lives of many Americans began to be connected to the practices of ESC research. Furthermore, it was argued that federal funding for ESC research would also help sustain US leadership in science and technology. 'Federal funding is probably required in order for the United States to sustain a leadership in this increasingly important area of research' (National Bioethics Advisory Commission, 2000, p. 60). There was no question, so the policy narrative went, that ESC research raised all sorts of ethical and moral questions. But one needed to consider that, according to the current consensus in the scientific community, ESCs were not embryos. Provided the adequate precautions were taken, research could continue in a manner that might lead to substantial benefits for US society and sustain scientific leadership of the United States in the world. This incorporation of important themes of US political identity into the emerging policy narrative was no doubt a powerful strategy to explain and justify the government's plan to begin with the financial support of a highly controversial line of research.

Despite its many words, the new policy narrative also had its silences. The most striking omission was the absence of any attempt to regulate private research – except for a NBAC recommendation to follow voluntarily the recommendations set out in the new guidelines (National Bioethics Advisory Commission, 2000, p. 79). Evidently, the rights-focused policy discourse that had dominated the US debates from abortion to embryo research since the 1970s continued to considerably define the logic of regulatory decision-making.

Thus, the NIH's policy narrative combined silences with detailed arguments, in a complicated and heterogeneous process of signification generated with the help of medical definitions such as what constitutes an embryo, and with references to important themes of the US political metanarrative, such as the duty to ensure future US scientific leadership and to protect the health of as many Americans as possible. Using this interpretation, ESC research became an extended promise concerning the future of both medical research and America, an expression of a particular image of US political identity. Government support for ESC research was equated with reason and an orientation towards progress and modernity, but done with careful deliberation. With these arguments, technological representations of the human body and interpretations of politics and economy, the space of policymaking began to be structured, policy actors came into the situation to position themselves in the policymaking process, alliances between institutions and actors became possible and, through the power of hegemonic definitions, political and scientific realities gradually began to take shape. With the NIH draft guidelines, it seemed that a decisive shift had been made in dealing with embryo research in the United States.

To be sure, the many opponents of ESC research had attacked this line of argument based on a counternarrative related to the abortion debate and cloning that attempted a different structuring of the policy space. The focus of this counternarrative was on the representation of embryos as human beings (not a formation of cells), their rights and the duty of the law to protect them. Furthermore, adult stem cell research and other research strategies were portrayed as viable alternatives to hESC research.

As a result of this strategy, and despite the many critical voices, the NIH were in a position to go ahead with its interpretation of the world of hESC research. The 'success' of the policy narrative articulated in the NIH guidelines was based on the hegemonic strategy to create what could be broadly perceived as 'a reasonable social compromise' in which all enlightened Americans should be able to recognize themselves. This increasingly dominant discursive structuring had a crucial impact on the meaning of the arguments and actors that interacted in the political struggle. The critics' discourse of morality was replaced by a discourse of 'utility-driven policymaking' with the goal of finding a 'common ground'. Support of ESC research was equated with reason, opposition with lack of reason. As the director of Yale University's medical school residency training programme put it: 'It is time all sensible people with an interest in humanity's future banded together to prevent misguided individuals from preventing right decisions' (Kapadia, 2000). The conflict about hESCs seemed to reflect a tension between two poles: 'the Americans', wedded to science, progress and modernity, and their adversaries, the forces associated with religion, ignorance and backwardness. The passing of the NIH guidelines for research using human pluripotent stem cells initiated federally supported embryo research in the United States – and indicated the agency's trust in the discursive power of its policy narrative to define the nature of a newly emerging scientific-political field.

Stem cells between Clinton and Bush

After a long and emotional battle, the new NIH guidelines came into force in late August 2000. They reflected a new policy dramaturgy and an attempt towards a proactive approach in stem cell research in a highly unfavorable political-regulatory context. This context included almost total lack of regulation of reproductive medicine, boundaries of life issues arising periodically and a polity characterized by a private/public divide that incorporated in these issues an irreconcilably divided American public. As we shall see, the new policy dramaturgy that attempted to reshape the meaning of hESC research from an attack on humanity to a contribution of medical progress and thus disconnect it from other boundaries of life issues was only temporarily successful. With the unfortunate combination of the NIH being both a supporter and a regulator of stem cell research, in the absence of trusted regulatory institutions, in the context of a notoriously unregulated constellation in reproductive medicine, the continually reemerging cloning topic and fierce criticism from the religious Right, the Bush administration soon found itself under immense political pressure to reconsider the policy strategy of the previous administration.

After President Bush took office in 2001, rumours immediately surfaced that the new president would change the government's stem cell policy and adopt a more restrictive approach towards funding and regulation (*Los Angeles Times*, 2001). In a nationally televised speech on 9 August 2001, Bush outlined the government's new policy in a little more than 10 minutes. In its essence, Bush announced that in the future federal funds would only be used for research on existing stem cell lines that were derived with the informed consent of the donors and from excess embryos created solely for reproductive purposes (White House Fact Sheet, 2001).

It would be a mistake to interpret President Bush's 9 August decision as a reversal of the Clinton administration's approach. It was in many ways consistent with the previous policy. The Clinton administration's approach towards embryo and fetal research combined support for experimental medical research with an attitude of caution and elements of restriction. Bush's decision fits well into this pattern. Certainly new rules were imposed on the conduct of ESC work, but federally supported research was nevertheless allowed to continue (National Institutes of Health, 2001b). In fact, the Bush decision was instrumental in broadening the endorsement for stem cell research and continued the earlier strategy of reabsorbing discourses of polarity into a system of 'legitimate differences'. In the logic of the Bush decision, the central conflict in stem cell policy was not between medical research and the rights of the unborn but between those who wanted to develop new cell lines and those who were satisfied to work with the already existing lines of ESCs. On a symbolic level, the new regulations offered to protect simultaneously the rights of the unborn and the freedom of research, and to foster medical progress in America.

Because of the new strategy, the initially well-defined boundary between the supporters and the critics of stem cell research began to blur, and the base of its social support was broadened. Whereas a number of the critics of embryo research, such as the Catholic Church, condemned President Bush's decision as insufficient to protect unborn life, many social conservatives from the National Right to Life Committee (NRLC) to Jerry Falwell voiced support or at least understanding for the new policy (Sack and Niebuhr, 2001).

The discussion continued as to whether a sufficient number of stem cell lines existed to continue with research. The WARF said it had filed a lawsuit against Geron Corporation on 13 August 2001 in the federal court to ensure broad research access to the five stem cell lines developed by researcher James Thomson. Needless to say, Bush's decision only affected the federal government's financial support strategy for hESCs. ESCs research in the private sector continued to be largely untouched by any federal regulations. The WARF had licensed certain rights for hESCs to Geron since 1999, in particular, exclusive rights for converting stem cells into six human cell types, those of liver, muscle, nerve, pancreas, blood and bone. The foundation sued Geron over the right to develop further rights, since there were more than 260 different types of cell lines in the human body (Wade, 2001). However, Bush's decision undoubtedly strengthened the position of both the University of Wisconsin and Geron, as the United States was the only country to have granted a patent on hESCs.

In April 2002 NRLC started a new phase in the struggle about stem cell research. In the past, critics of stem cell research had linked the research to abortion and 'killing babies', but critics now began to embrace a much broader agenda, the 'protection of the sanctity of life', of human civilization 'as we know it'. Whereas the British government seemed to be firmly in control of 'threats' such as human cloning, the American government seemed to be floundering, a situation that critics seized upon. After the stabilization of stem cell policymaking under the Clinton administration and attempts to separate human cloning and abortion from hESC research, the 'cloning threat' re-entered the political stage. The paradox of the public/private divide of reproductive medicine research in the United States had created new entry points for the critics of hESC research. Again, the cloning topic was powerfully brought into the controversy, and the argumentation blurred issues of questions of reproductive and therapeutic cloning with hESC research. The accusations that the private sector was a 'Reproductive Wild West' seemed to receive support by the disclosure by Advanced Cell Technology (ACT) in November 2001 that it had created the first cloned human embryos, which survived for several hours. The results of this research were published in an electronic journal (Cibelli et al., 2001).

Another important event was the announcement from a company called Clonaid that the first cloned human had been born, a seven-pound baby girl nicknamed Eve, born on 26 December 2002 outside the United States. The attempts to redefine the stem cell topic were followed by myriad legislative initiatives, all of them quickly counteracted by other initiatives designed to offset these attempted restrictions of research. In addition, in 2001 several lawsuits were filed relating to stem cell research issues.

The ACT announcement partially triggered the first project of the newly created President's Council of Bioethics, whose creation had been announced in President Bush's speech in August 2001. At its first meeting, Dr Kass announced that the first topic to be addressed by the Council would be human cloning. In its final report, *Human Cloning and Human Dignity: An Ethical Inquiry*, a majority of the panel had voted for a four-year ban on cloning for biomedical purposes. (President's Council on Bioethics, 2002).

These developments on the federal level were countred by a series of initiatives on the state level culminating in California's proposition 71, a referendum in 2004 that gained a majority securing 3 billion US-D funding for stem cell research (including hESC research) in California for a period of 10 years.

During this time, supporters of stem cell research such as patient groups and scientific organizations continued to articulate their strong endorsement of stem cell research. But with stem cell research beginning to be associated with cloning, a broad alliance of support for restricting the science had emerged. In the middle of 2003, the debate seemed to be paralysed, with critics and supporters simultaneously making advances in translating their positions into policies. Just as national abortion policies since 1973 had been undermined by state policies, the federal government's approach towards stem cell research seemed now to experience a similar development.

This constellation was far from a policy dramaturgy in which a proactive, deliberative style of policymaking seemed to be in control of hESC research. While the Bush administration approach was hardly enthusiastic about supporting hESC research, it still constituted a continuation of the course of the Clinton administration. With a 'Wild West constellation' in the private sector that continually issued news from the brave new world of reproductive medicine causing fright and disorientation among American citizens, along with the general regulatory vacuum in embryo and IVF research and the continued absence of trusted regulatory institutions in hESC research, the controversy continued around broad boundaries and culture of life topics, which hindered the further development of the NIH approach launched in 2000.

5
From Dolly to Therapies?
Stem Cell Regulations in the
Making II – Germany, Italy,
Japan and South Korea

Defending nations, dignity and embryos

In this chapter we will further elaborate our central argument that the shaping of hESC policies is certainly determined by numerous factors including history, culture and religion, but that discursive memory, narrativizations and decisions to create particular policy dramaturgies clearly play a very important role in shaping regulatory outcomes. After having discussed the two leading centres of hESC research, we will now move to a set of countries that offer striking differences in their approach towards stem cell regulation, Germany, Italy, Japan and South Korea.

Germany

As mentioned in Chapter 3, from the very beginning of the debates on Dolly and hESC research, Germany was in a special situation. In most other countries, Dolly propelled the question of a ban on reproductive cloning and the development of regulatory measures dealing with hESC research onto the political agenda, but in Germany the situation was just the opposite: with the Embryo Protection Act, a sweeping ban had already been imposed on embryo research, including any manipulation of embryos after fertilization with any purpose other than resulting in a pregnancy. By definition, with the Embryo Protection Act mixing issues of embryo research, abortion and artificial reproduction, any regulatory developments dealing with hESC research entered the diffuse field of the politics of defining or redefining the boundaries of life.

As soon as ESCs began to emerge as a topic in Germany, they were linked not only to Nazi eugenics but also to questions of human rights, the collective good and the 'political nature' of Germany. In the gradually dominant policy narrative, ESCs were interpreted as part of an important development, which, however, should not be overrated. German history and the resulting responsibility – so the narrative went – made it crucial for Germany to proceed with extreme caution. In this increasingly hegemonic construction of the political space, opposing views that argued for the necessity of catching up with international scientific developments, like those of the Deutsche Forschungsgemeindschaft (DFG), were not simply negated. Instead, aspects of these positions were integrated into the dominant narrative, in particular through reference to the promise of adult stem cell research and the possibility of conducting research with aborted embryos (Catenhusen, 2000, p. 124; *TAZ*, 2000; Frankel, 2000).

Hence, in Germany the field of hESC research was discursively constructed between the pole of medical research that was connected to German health and modernity and the pole of a broadly conceived notion of human rights, associated with the collective good and 'learning from history'. In other words, the staging of the ESC research conflict was not construed as a tension between research and backwardness that coded reason against lack of reason – as was the case in the polarized constellation in the United States – but as a tension between research and human rights. This particular discursive construction was hardly a logical and direct outcome of historical development or religious preferences but was an interpretive intervention shaped in a process of exchange with significant impact on evolving policy negotiations.

This approach found its most dramatic expression in a widely publicized speech by German President Johannes Rau, who cautioned that in the light of German history, technical feasibility in medical research should never be an argument to relativize ethical concerns (Rau, 2001b). Unlike the dominant US criticism of ESC research, which was framed as an expression of religious opposition and narrow mindedness, in Germany's 'discourse of morality' the rejection of ESC research became synonymous with a number of broadly shared social and moral goals, such as the integration of disabled people into society, the protection of the interests of children and women, and the preservation of a 'good' and ethical German society. There was a price to be paid for this 'Sonderweg' such as a disadvantage in medical progress – but it was worth it.

This framing of the critique of ESC research created a particular choreography of the evolving conflict: it allowed a variety of different groups to rally behind the rejection of this type of research and made possible the particular German dramaturgy of hESC regulation. As a result, opposition to hESC research was not confined to mostly religious groups and arguments (such as in the United States) but reached from the churches to the Socialist Party, the Conservative Party, the Greens, the German president and many other positions and groups. While a 'rights of the embryo' argument might have immediately split the opposition against stem cell research, the mobilization and combination of a plurality of themes connecting German history with human rights topics discursively created a broad alliance of groups and actors critical of ESCs. In this constellation, any plan to reform the Embryo Research Protection Act came across as a highly questionable political project. The dominant 'discourse of morality' left little space for compromise, utility considerations and 'normal' policymaking – only for the 'right decision': to reject any notion of reforming the Embryo Protection Act and to prohibit hESC research. Whereas a number of individual medical scientists advocated a pragmatic approach towards hESC research that would have permitted research under special provisions, for almost 3 years none of the potential architects for a counternarrative about stem cells, such as the Ministry of Health, the DFG or the National Chamber of Doctors, risked moving forward with a supportive interpretation of stem cell research and a proposal for action. The name of the game was non-decision-making, a drama without much action.

Towards the end of the year 2000, reports came from the United Kingdom that the British parliament had passed legislation legalizing therapeutic cloning. In Germany, a discussion immediately followed about the ethical aspects of this decision, but more prominently about its economic implications and the perils of the cautious German approach in this field. German Chancellor Gerhard Schröder soon wrote a much publicized and quoted article 'in which he forcefully demanded an adjustment of the German approach towards genetic engineering and new medical developments'. But even Schröder cautioned that the potentials of adult stem cell research needed to be explored first. However, his invitation to a 'politics without blinkers', as he expressed it, became a rather notorious statement in the gradually developing German 'bioethics discourse' and was widely interpreted as suggesting a bolder and less restrictive approach towards biotechnology and reproductive medicine (Schmiese, 2000).

Schröder's speech was countered by a widely received speech about a month later by German President Johannes Rau given at the Deutsche Bundestag to memorialize the victims of National Socialism. In his speech, Rau described the devastating impact of National Socialism. Towards the end of the speech, Rau discussed the relationship between science and National Socialism and connected these experiences to today's challenges in biomedicine (Rau, 2001a).

Only towards the end of 2001 were demands put forward, such as those by the DFG to consider the importation of hESC lines, to end the de facto ban on ESC research in Germany and providing a 'quick solution' to the needs of German medical research (*TAZ*, 2001a). The DFG, the central German institution for the support of basic research in Germany, presented a bold plan for action: 'In the last two years ... major progress has been made in stem cell research. The DFG is of the opinion that now science has reached a level of development from which potential patients as well as scientists in Germany should not be any longer excluded.' The DFG plan recommended that a first step could be made through international collaboration, without any need to change the Embryo Protection Act. But, as the DFG put it, the simple importation of stem cell lines was not enough. In a further step, German scientists should have the opportunity to actively take part in the creation of hESC lines. This new option should be available for 5 years and then be reviewed. Reproductive and therapeutic cloning should not be permitted. Concluding the plan, the DFG stated: 'Against the background of the more recent German history the DFG is aware of the problem to use, if not produce, early human life for research purposes. The DFG is convinced that its recommendations correspond with our understanding of law and constitution, but also our model of man that takes into account the interests of scientific research and sick people' (Deutsche Forschungsgemeinschaft, 2001).

It was in this constellation that Johannes Rau gave another speech: 'Will everything turn out well? For progress on a human scale' (later also published as a book, Rau, 2001a). The speech drew intense attention and constituted a sweeping rejection of hESC research, combining in its treatment stem cell research, PGD and euthanasia: 'Progress on a human scale knows its values. The opposite to unlimited progress is not standstill or falling back. Who opposes progress at any cost, is not an enemy of progress. ... Life reminds us that we humans, despite all progress, remain finite beings. When we pretend that our possibilities have no limits, we ask too much from ourselves. It is then that we lose our human scale.' A few days later, the Protestant Church of Germany, referring to Rau's speech, unequivocally rejected hESC research, attacking the DFG's demand to import hESC lines as ethically highly questionable. While the Protestant Church in Germany had traditionally taken a progressive course, for example, expressing its understanding of abortion under certain circumstances, it clearly conveyed in its statement the need to take a clear and negative position towards embryo research (Evangelische Kirche Deutschlands, 2001). The Catholic Church was also clearly opposed to any form of hESC research in Germany (Deutsche Bioschofskonferenz, 2001). And the German Doctors' Association (Deutscher Ärztetag) passed a resolution at its annual convention that there was no need at this stage, as the DFG had demanded, to start hESC research in Germany and to reform the Embryo Protection Act (Deutscher Ärztetag, 2001, p. 7).

Within a few weeks a polarization had taken place in Germany that resulted in division of the political field between the government, represented by Chancellor Schröder, the FDP and representatives from SPD, CDU, and CSU, and the DFG, and a large section of the SPD, the CDU/CSU, a unified Green Party, the German Doctors' Association and the Protestant and the Catholic Churches. Ulla Schmidt, a new (SPD) Health Minister, had succeeded Andrea Fisher, the Green minister who had stepped back in the wake of the BSE scandal. Fischer was, as shown earlier, highly critical of hESC research, but with Schmidt, a consolidation within the SPD ministers in the cabinet had taken place. It was in this constellation that the governor of Nordrhein-Westfalen, Wolfgang Clement, announced that he would support the signing of a treaty between the University of Bonn, where Germany's most prominent stem cell researcher, Oliver Brüstle worked, and the University of Haifa. This announcement caused an uproar across parties, and the DFG, which planned to make a decision about importing human stem cells on 3 July, was asked by the CDU, the SPD and the Greens to delay its deliberations until the parliament had reached a resolution on the topic (TAZ, 2001b).

In June 2001 Chancellor Schröder launched the much-discussed National Ethikrat (National Ethics Council), an institution assigned the task of advising the chancellor on bioethical questions. The formation of the Council was clearly driven by the desire to create a location for deliberating sensitive issues such as embryo research, but its creation came relatively late, and it was soon criticized as being a tool of the SPD to legitimize the idea of medical progress. Whereas the United Kingdom had the HFEA, Germany was struggling to create an institution of authority that could be expected to gain the trust of the citizenry. Other than this council, no tangible efforts were made to connect with different publics such as stakeholders, patients or citizen groups, or even to hold a consultation of some sort, as was so typical for Britain and Japan (see following). The polarized debate operated at the level of the general public, and it seemed to be pressured by time constraints and the need to arrive at political conclusions quickly, but it is questionable if this particular dramaturgy was the best to reach a broadly satisfactory outcome.

Stem cells in the Bundestag

While 'the Chancellor's' Ethikrat was working to issue its opinion on stem cells before the end of 2001, the central stage of the German hESC controversy shifted to the Bundestag. The parliament's inquiry committee Law and Ethics in Modern Medicine, which had started working in May 2000, was developing a position on the ethical and sociopolitical implications of ESC research. The Enquete-Kommission had been charged with examining progress and problems associated with medicine and research practice, and the related ethical, constitutional, legal and political issues (Deutscher Bundestag, 2000, #80). On 21 November 2001, the Enquete-Kommission presented its second report (Deutscher Bundestag, 2001, #156), a massive, 150-page document that constituted a thorough examination of the scientific, legal and ethical implications of stem cell research. In its main conclusions, however, the commission clearly rejected a liberalization of the German Embryo Protection Act and also voted against the importation of hESCs (57–8). On 29 November 2001, the National Ethics Council finally announced its position in the stem cell debate. The Council's position statement, published on 20 December 2001, focused on the question of importing of hESCs and stated that 15 members of the Council had voted in favour of importation, with 10 voting against it (Nationaler Ethikrat, 2001). With these developments, the struggle over stem cells increasingly took on the air of a confrontation between the Chancellor and the parliament. German newspapers devoted a great deal of attention to what was broadly interpreted as a Council without legitimacy (*Die Zeit*, 2001b) – and to the German parliament, which mobilized all its resources across party lines to face up to this political challenge. In addition to the Chamber of Doctors, important patient groups and organizations for mentally and physically disabled patients spoke strongly against permitting the importation of hESCs to Germany (Brocke, 2002; Deutsche Alzheimergesellschaft, 2002).

On 30 January 2002, the German parliament began a debate that took more than 5 hours and that became the key event in the German debate about hESC research. Although Bundestag votes are usually very predictable, as German members of the parliament follow strict party discipline, in this case 'party discipline' was lifted, and in the weeks before the decisive parliamentary vote intense negotiations between members of the different parliamentary factions began (*Die Welt*, 2001;

Cimons, 2001). This symbolically added to the aura of the 'exceptional' nature of the hESC question, and elevated the parliament, usually closely following the direction of the government, into a location for decision-making on a key issue of humanity, the question of human dignity, and its protection by the state. Thus, a scenography had been created in which the central actors, the members of the parliament, were not simply deciding about the regulation of hESC research, but about the future of humanity and the state of the German nation. The formal topic of the parliamentary debate was the discussion of the second report on stem cell research, by the parliamentary inquiry commission Law and Ethics of Modern Medicine (Deutscher Bundestag, 14. Wahlperiode, January 2002a). Simultaneously, three motions were filed by three groups of representatives, which proposed different models for dealing with the question of importing hESC lines from abroad. The Bundestag was to vote on these motions, which were not bills or any other sort of regulations but outlined scenarios for future action ranging from permitting the importation of hESC lines within the framework of existing legislation to the adoption of a new law banning importation. The motion by Wodak, Kues, Knoche et al., a joint Green, CDU/CSU, SPD, PDS effort, was labelled 'Protection of Human Dignity in face of biomedical opportunities: no import of embryonic stem cells' and demanded a strict ban of hESC research and the importation of stem cell lines from abroad. It argued that any attempt to assign a different moral status to embryos created outside of Germany than to embryos created inside, would lead to a 'Doppelmoral' (double moral). Furthermore, the only reason the importation was not banned in Germany was that at the time the Embryo Protection Act was passed, the legislation was in no position to anticipate later scientific developments. Therefore, one could not deduce the legality of the issue from the lack of penal regulations. Women would be used as the producers of the 'raw materials' for stem cell research. Importing stem cell lines from abroad would soon turn out to be inefficient, and eventually there would be pressure to conduct embryo research in Germany. The immune rejection problem associated with hESCs from donors would put pressure to move towards therapeutic cloning (Deutscher Bundestag, Antrag der Abgeordenten Dr Wolfgang Wodarg, Dr Hermann Kues, Monika Knoche et Al., 14. Wahlperiode, Drucksache 14/8101, 29 January 2002, pp. 1–6).

On the other side was the motion by Ulrike Flach et al., an FPD-dominated initiative joined by CDU/CSU and some SPD members, with the name 'Responsible Research with embryonic stem cells for an ethically sound medicine'. This motion argued that no reason existed to restrict importing ESCs, that stem cell research gives hope to many very sick people, and thus within a tight system of regulation, research should proceed. Furthermore, this motion also demanded that if the anticipated success did not come with imported stem cell lines, the German Embryo Protection Act should be 'further developed', meaning, as would become clear during the parliamentary debate, that the act should be rewritten so ESCs could also be obtained in Germany (Deutscher Bundestag, Antrag der Abgeordneten Ulrike Flach, Katharina Reiche, Peter Hintze et al., 14. Wahlperiode, Drucksache 14/8103, 29 January 2002, pp. 1–3).

Finally, there was motion 14/8102, with the title 'No utilizing embryo research: prohibit the import of embryonic stem cells and admit it under certain circumstances'. In this 'motion of compromise', sponsored, among others, by the former Green Minister Andrea Fischer, the chairperson of the Law and Ethics of Modern Medicine Inquiry Commission (Renesse, SPD), and the CDU/CSU Health Expert Horst Seehofer, a proposed law prohibited the importation of ESCs for private and public purposes. However, exceptions could be made if (1) alternative research methods did not seem to be promising, (2) the imported stem cells were generated prior to the date of the decision about the motion, (3) the embryo donors gave their consent to use the stem lines, (4) the excellence of the research proposal utilizing ESCs could be proven, (5) the ethical acceptability of the research was studied by a central ethics commission (6) and a new authority was set up in charge of the necessary procedures. Only under these circumstances would the new law permit ESC research, and only under these circumstances could it be guaranteed that no new embryos would be killed to make research possible (Deutscher Bundestag, Antrag Dr Maria Böhmer, Magot von Renesse, Andrea Fischer et al., 14. Wahlperiode, Drucksache 14/8102, 29 January 2002, pp. 1–5).

With these three motions a completely unusual situation had been created in the Bundestag. Not only was party discipline lifted, but also three different proposals for action had been drafted, each supported by members from several parties. A policy drama had been set in motion whose plot had significantly changed the usual rules of the political game. Speculations abounded which proposal would get the majority. While it seemed relatively clear that the most liberal proposal had no chance of passage, it was completely open which of the two competing proposals would carry the majority of votes.

After a long, almost five-hour-long discussion, two votes were necessary. In the first vote the ban-motion received 263 votes, versus 226 votes for the compromise vote, and 106 votes for the pro-vote (21239). With this result, no motion had gained a majority. In a second vote, the compromise vote gained 339 votes versus 266 votes for the ban-vote. This result was widely interpreted as a 'victory' of the moderate forces, but also of Chancellor Schröder (Polke-Majewski, 2002). Soon after the vote, the chairman of the Conference of Catholic Bishops, Cardinal Lehrmann, and the chairman of the Council of the Protestant Church, Manfred Kock, stated: 'Thus it will be also possible in Germany to experiment with embryonic stem cells which were gained by killing embryos. By this decision the right to live and the unlimited protection of humans from the beginning of conception is not guaranteed anymore' (Deutsche Bioschofskonferenz, 2002).

The decision of the Bundestag was seen not only as highly controversial but also as paradoxical. A narrow majority of the representatives had voted in support of a provision that allowed the importation of hESC lines created before 30 January 2002 for research purposes. But the same parliamentary decision declared the importation of hESCs as illegal. Only under special circumstances could exceptions be made, and in such cases, a number of restrictions were imposed. For example, those applying to import hESCs were requested to document that for the planned research, alternative strategies would not yield the same expected results to from the usage of hESCs. Finally, the Embryo Protection Act remained untouched by the legislators. The parliamentary decision from 30 January 2000 was seen as filling a gap in the Embryo Protection Act created by the advancement of science, rather than reversing it (Polke-Majewski, 2002). Apparently, even in early 2002, the option to change the established regulatory regime that protected the embryo for the broad majority of policy actors was out of question. A well-defined boundary continued to separate the supporters of research from the advocates of human rights. In this constellation, the importing of hESCs from abroad constituted a 'quick solution' for the needs of medical research, but the interpretation continued to dominate the political debate that there is fundamental conflict between the conduct of embryo research and the principles of human rights. The vote from 30 January 2002 was a vote of principle, but not yet a legislative vote. A new law regulating the importation of ESCs still had to be devised.

On 28 February 2002, a bill drafted by a group of members of the parliament largely synonymous with the group that had supported the 'compromise motion' in January was sent to the relevant parliamentary committees. It insisted that no stem cells derived before 1 January 2002 could be utilized. With a positive vote by the Bundestag (*TAZ*, 2002), the new law went into force on 1 July 2002.

Thus, 'even' in Germany hESC research could go ahead after 2002. From its inception, developing policy regulations for hESC research were destined to be built around the existing Embryo Protection Act and thus connected research on early embryos with an extensive chain of associations from National Socialism and the need to protect human dignity to many other questions of the boundaries of human life. With the interpretation shared by many important actors in Germany that any kind of legalization of hESC research constituted an attack on the act and human dignity, the development of a proactive strategy to deal with the regulation of hESC research turned into a towering task. A policy dramaturgy had been set in motion in which defenders and relativists of human dignity were positioned to struggle with each other, and the regulation of hESC research as such almost became a secondary issue.

Italy

As we discussed in Chapter 3, the regulatory vacuum in reproductive medicine in Italy had increasingly become a political liability for many parties, and this burden eventually resulted in a radical turnaround of the regulatory environment. The National Committee for Bioethics (Comitato Nazionale per la Bioetica, CNB) started to ponder hESC research in April 2000 and approved a final report on this subject at the end of October 2000 (Comitato Nazionale per la Bioetica, 2000). In this report, the CNB approved the derivation and usage of stem cells from adult tissues and aborted fetuses, but the Committee's members were split on the question of the ethical permissibility of hESC research. All members agreed that embryos should not be produced for the purpose of research, but they did not achieve a consensus on how to proceed with 'surplus' embryos, that is, embryos initially produced in the course of fertility treatments but no longer needed for that purpose. While the CNB was still working on its final report, Health Minister Umberto Veronesi gave an ad hoc commission the task of providing evidence on the science and ethics of hESC research. This commission was composed of 25 members, chaired by Nobel Laureate Renato Dulbecco, and therefore known as 'Commissione Dulbecco'. The final report of the commission was published in December 2000 (Ministero della Sanità, 2003 [2000]). Although the members of the Dulbecco Commission agreed on the scientific stakes of hESC research, they were split on the ethics. All members agreed that embryos should not be produced for the purpose of research, but they could not find a common ground on the ethical legitimacy of using surplus embryos for research. However, the commission was finally able to reach a consensus on how to bypass this conflict. It presented a sophisticated argument on SCNT, arguing that the technology could be altered in such a way as to inhibit the creation of human embryos, thus providing a technical answer to ethical problems and – in the words of the Ministry of Health press release – an 'Italian way to therapeutic cloning' (Ministero della Sanità, 2000). Nonetheless it did not pave the way for a future policy narrative, on the contrary. In the first weeks after publication of the report, questions were raised about the scientific feasibility of the 'Italian way towards therapeutic cloning' as well as questions on whether SCNT would 'really not give rise to human embryos' – after all, 'Dolly the sheep was birthed that way'. As it turned out, in the long term, the experience of the Dulbecco Commission and with it the attempt to order stem cell research through their staging of expert committees proved to be the initiative of a single Ministry, that was anchored neither in a larger political context nor in Italian political culture. Subsequently, stem cells, clones and embryos continued to be an unresolved topic in Italy's sociopolitical reality.

When the law 40/2004 was published in Italy's *Official Gazette* on 24 February 2004 (Repubblica Italiana, 2004), an enduring and rocky process of legislation had ended. After 7 years of waiting, the very enactment of the law was celebrated as an achievement, ending Italy's 'wild west' of reproduction research. Throughout the years of puzzling over rights and obligations in an age of new reproductive choices, a particular 'policy narrative' had emerged. This narrative was inscribed into the law 40/2004, and applied this system of meaning to all of Italian society. The narrative's protagonist was the early human embryo. It was framed neither as a cluster of cells nor as *any* human being, but as the first embodiment of a small, innocent child and future citizen of the Italian Republic. The child had rights that were threatened, so the narrative went, by the selfishness of its future parents and by the drive for financial gains of maverick doctors. The state must therefore safeguard the embryo's rights. Indeed, the first article of 40/2004 solemnly pronounced the basic aim of the law was to 'ensure the rights of every involved subject, including [the rights of] the conceived (concepito)' (Repubblica Italiana, 2004, Art. 1, Sec. 1).

Among the rights of the 'conceived' was the right not to be produced for any other purpose than to become a child. This implied that no more than the 'strictly necessary' number and – at any rate – no more than three embryos may be produced, that all produced embryos must be transferred, and that no embryo may be frozen.[1]

In short, a policy scenography had emerged that reacted to a constellation of confusion in the field of reproductive medicine in Italy, but focused even less on hESC research as such, which had become a sub-story in a much more complicated drama. The rise of this particular policy narrative did not take place in a void. But it would be misleading to conceptualize the law as Catholic rather than Italian. Although Catholic discourse on the subject played a role, Italian parliamentarians were concerned with the makeup of the Italian collective body, with the care of its future citizens and with the (familiar) environment in which they lived. They struggled with how to guarantee a proper re-creation of the nation and how to reorder the micro 'cells' – the families – on which the 'molar' Italian body rests. The fear of the disruptive consequences of IVF on the Italian family was particularly important. Some parliamentarians appeared anxious to keep the permitted range of techniques of assisted reproduction as close to 'nature' as possible and to ensure that artificial techniques were deployed only to give 'nature' a little hand, reaffirm-ing the traditional boundaries of the family rather than challenging or disrupting them.

The discursive reordering of reproduction on the stage of the Italian parliament was shaped by various narratives, by Catholic discourse, the care of the family and its future citizens, and a care for 'nature'. Although anchored in the Italian discursive economy, this reordering was not shared by all members of the Italian parliament or by all of Italian society. When Italian legislators began to struggle to understand and tame Italy's 'Wild West of Reproduction' on the eve of the birth of Dolly in 1997, they worried about how to restructure the 'reproduc-tion of the nation' and fix the fluctuating definition of families and parenthood rather than concern themselves with science or stem cell research. In other words, the prospect of regulating hESC research through proactive legislation had vanished in a constellation where the boundaries of life, family and the Italian nation determined the politi-cal course of action.

Towards the referendum

In fact, it is difficult to find references to stem cell research or the implications of cloning or embryo research for science and the genera-tion of new truths in the thousands of pages of the transcripts of the debates on law 40/2004. Once the law was enacted, however, it had important implications for hESC and cloning research. Stem cell research was not forbidden, but Italian researchers could only gain access to hESCs through importing them (Pasotti and Stafford, 2006). The 'Association Luca Coscioni' began to mobilize against the legislation. Named after its founder, Luca Coscioni, who suffered from amyotrophic lateral sclerosis

(ALS), a disease that figures prominently among the potential targets of hESC and cloning research, this organization was a transversal association that involved a broad range of sociopolitical actors, such as scientists, physicians, infertile couples and patients affected by chronic diseases. In spring 2004, the Luca Coscioni Association started collecting signatures for a petition for a referendum to overturn the entire law and add four requests aimed to partially cancel the law. The first sought to cancel all passages of the law that hampered hESC and cloning research in Italy, such as the ban on embryo research, on the production of 'surplus' embryos and on SCNT, in order to enable 'new cures for diseases such as Alzheimer's, Parkinson, sclerosis, diabetes, heart diseases and tumours'. The second petition sought to cancel the ban on embryo freezing and all provisions that outlawed preimplantation genetic diagnosis. The third sought to reaffirm '[women's] auto determination and the protection of women's health' and to abrogate the passage of the part of the law that framed the early human embryo as having equivalent rights as its future parents. Finally, the fourth petition sought to reintroduce donor insemination to the range of permissible practices in Italian laboratories. Overall, more than a million signatures were collected. The Constitutional Court ruled on the admissibility of the five petitions in January 2005. It did not admit the request for the abrogation of the entire law, arguing that a law was constitutionally necessary, but it gave permission for the four 'partial' referenda.

Article 75 of the Italian Constitution has arranged for the possibility for the 'sovereign', the 'Italian people', to cancel the work of its representatives through an abrogative referendum (*referendum abrogativo*). It permits popular referenda in that the Italian electorate can give direct expression of its will about a specific law *after* the law's approval. Thus, a referendum can nullify an existing law, either partially or completely. But this tool of direct democracy also has some pitfalls: for a referendum to be valid a minimum of 50% +1 of the Italian electorate must cast its vote.

From early 2005 to the 12th and 13th of June, when the referenda took place, stem cells, clones and embryos ceased to be just the stuff of expert committees, parliamentary debates or the discussions of activists. They became the object of electoral campaigns, newspaper headlines, television shows, and entered the territory of the 'general Italian public'.

The debates were organized by the emergence of two narratives. The first narrative – which we could call the '(vital) rights and freedom' alliance – focused on human rights and freedoms: the freedom of reproduction, the freedom from religion, women's rights over their bodies, the freedom to make one's own decisions unhindered by someone else's ethical choice or national or governmental morals, the freedom of research, the right to obtain cures and to live a healthy life, and the right to have – at least hope for – potential therapies in a not-too-distant future. Its reference point was the 'human being' as a biological citizen whose genetic or somatic corporeality, vulnerability and suffering endows the person with vital human rights, which the state as a good shepherd must not interfere with but, in fact, must foster. The second narrative, too, referred to 'human beings', radicalizing this category to include the 'life' of fertilized oocytes and embryos. As this 'life' is 'one of us', so the argument went, it needs protection, in extremis, at the expense of the rights and liberties of the qualified life of the 'citizen'. Both alliances performed their own realities, truths and – indeed – visions of society; but they were not equal: one of them was more successful. Considering the turnout of the referenda, it was obviously the latter. As three-fourths of the electorate chose to stay home, it was a merely statistical detail that between 78.2 per cent (for the legalization of gamete donation) and 89.2 per cent (for the legalization of hESC research) decided to vote 'yes'. The referendum had failed, and with it the attempts to partially revert the 2004 law. The paradoxical absence of discernible public policy strategy in hESC research and of trustworthy institutions that could regulate hESC research provided a context in which questioning the 40/2004 embryo protection law and deregulating reproductive medicine was not seen as an acceptable option.

Japan

As in other countries, Japan's swift reaction to the question of reproductive cloning was not equal to the shaping of a regulatory framework for hESC research. However, what evolved gradually over the years after 1998 was an extremely cautious approach towards hESC research that years later would provoke harsh criticism from Japan's leading stem cell researcher, Norio Nakatsuji, who condemned the system of regulation as irrational and damaging to stem cell research in Japan (Nakatsuji, 2007).

Japan is not like the United States or Germany, where a subject like stem cell research can quickly turn into a hugely socially divisive issue and 'get out of control', but it was apparent to Japanese policymakers that hESC research was a highly sensitive topic. Because of the nature of Buddhist religion, which is not centrally but individually organized, it is difficult to identify anything like an 'official Buddhist position' on stem cell research (Schlieter, 2005). Nonetheless questions such as abortion or dealing with embryos clearly constitute potentially controversial issues in Japanese culture (LaFleur, 1992). At the same time, the virtual absence of regulations in the field of reproductive medicine make any policy process in this arena especially volatile (Mayumi, 2006).

The policy scenography of hESC regulation in Japan involved different locales, the interaction with various publics and a display of controversy not typical for the Japanese polity. In December 1998 the Bioethics Committee of the Council for Science and Technology had established a special subcommittee on the Evaluation of Research on hESCs, which issued a report in March 2000 with a recommendation that embryos used in research should be supernumerous embryos from IVF research. Based on this report, the government issued its hESC Research guidelines in September 2001. They stated that the Institutional Review Board (IRB) would review the protocols of both the donating medial facilities that store human fertilized embryos and the research institute that conducts the derivation of hESCs. Additionally, the Ministry of Education, Culture, Sports, Science and Technology (MEXT) would examine the IRB's review. Before the guidelines were issued, public comments were collected from 17 February to 19 March 2001, with the idea of incorporating them into the final version of the guidelines (Kazuto, 2005, pp. 371–2). While hESC lines were imported initially from abroad, the development of stem cell lines in Japan seemed to concentrate on Norio Nakatsuji's laboratory in Kyoto, which provided the other Japanese hESC researchers with stem cell lines. Overall, a strong determination developed among Japanese regulators not only to closely supervise hESC research but also to limit the use of fertilized embryos as much as possible, mainly by tacitly designating one laboratory, Nakatsuji's lab, as the source for hESC lines in Japan (Interview, Kyoto, 8 October 2004). Another highly controversial topic was the question of SCNT. When the law banning reproductive cloning was enacted in 2001, it stated that the questions of the creation of embryos for research purposes should be discussed in the committee of the Bioethics Investigation Panel in the Council for Science and Technology (Kazuto, 2005, pp. 375–6). The panel had more than 20 members from medicine, biology, humanities and social sciences and held 30 meetings between 2001 and 2004. It also organized public workshops in Kobe and Tokyo, but no consensus could be reached (Kazuto, 2005, p. 376), a situation highly unusual in Japanese politics, and as a result, the topic remained undecided. From its inception, Japanese hESC research was guided by a cautious, relatively open proactive strategy that combined tight regulatory measures with an approach of gradually opening up the possibility for stem cell research, which placed Japan in the regulatory spectrum somewhere between Germany and the United Kingdom.

Korean stem cell politics under the shadow of the Hwang Affair[2]

While the case of Japan seems to question the argument that countries with Buddhist religion tend to be permissive about hESC research, the relatively non-restrictive regulation of hESC research in Korea raises the question again of the relationship between Buddhism and regulation of stem cell research. But the Korean hESC strategy is less a reflection of Asian culture than of failed science governance that came at high costs for the Korean scientific community. This book is not the place to discuss the highly complicated rise and fall of Seoul National University Professor Hwang Woo-Suk; nevertheless the failure to establish a regulatory framework promptly and proactively to deal with hESC research in Korea in many respects is related to the misled attempt of powerful groups in Korean politics to create a favourable political-regulatory context for Hwang, who within a brief amount of time had become the leader of Korea's attempt to jumpstart its leap into international biotechnology leadership.

Already in 1994, seven South Korean government ministries had signed the 'Biotech 2000' plan with the goal of making Korea one of the top seven biotechnology players worldwide. This plan articulated a shift from the technology-learning paradigm to a technology-creation paradigm, which began to take shape in early 2000. Challenged by the Asian economic crisis of the late1990s, biotechnology and the newly emerging stem cell and cloning techniques were possible candidates for international leadership in a new, hot field of science and technology. During the mid-1990s, when in the United States and the United Kingdom researchers began to pioneer cloning and stem cell techniques, Korean researchers began to move quickly into the newly emerging field. In particular, it was the work of Hwang Woo-Suk, a Korean pioneer in animal cloning at the College of Veterinary Medicine, and his collaboration with Shin Yong Moon at the Medical Research Centre from the late 1990s at Seoul National University that gained the international limelight in 2004 and 2005. With the publication of what was regarded as two landmark papers on SCNT research in *Science* in 2004 and 2005, and the launching of the World Stem Cell Hub in November 2005, the success story of Korean stem cell research seemed to have reached its peak. A policy narrative became increasingly dominant to focus on Hwang's research as the pride of Korea.

Key policymakers seemed to do everything to avoid putting any 'regulatory hurdles' in the way of Professor Hwang (Interview, Seoul, 1 November 2005). The slow pace of regulatory policy development in the hESC and cloning field was a reflection of this approach, which together with widely unregulated reproductive medicine had created a regulatory vacuum that provided the environment for Hwang's research fabrications and ethical misconduct, with a tremendously negative impact on the entire South Korean scientific community.

As shown in the previous chapter, the broad debate on the issue of cloning gave little evidence that Buddhist culture somehow provided a shield for raising ethical questions with respect to embryo research. After the birth of Dolly, cloning was a widely and critically debated topic in Korea. In phase one of the Korean policy scenography, hESC research was staged as a controversial topic with many different actors positioned to speak up and develop a role in the decision-making process. First, the Ministry of Health and Welfare and the Ministry of Science and Technology, respectively, had started to take legal steps. In December 2000 the Ministry of Health and Welfare announced a Life Science and Health Safety and Ethics Bill. This bill suggested that human embryonic cloning as well as human cloning were to be forbidden. However, it did not evolve into further governmental action because of opposition backed by scientists and industrialists. The Ministry of Science and Technology prepared an autonomous bill by initializing a bioethics advisory committee as a consultation organization for the Ministry of Science and Technology. The Bioethics Advisory Committee, composed of 20 specialists from sciences, social humanities and NGOs appointed by Minister of Science and Technology, met 18 times from November 2000 to May 2001, and it announced a Bioethics basic bill along with its activity report. According to the report prepared by the committee, human embryonic cloning should be prohibited, whereas research on surplus embryos should be only partially allowed. Upon announcement of the report, scientists and industrialists intensely opposed the bill just as they did the bill proposed by the Ministry of Health and Welfare. Immediately after the public hearing, approximately 300 biologists and 13 scholars declared their opposition to the proposal. They claimed the proposal was too restrictive and could lower the international competitiveness of Korean biotechnology. The Ministry of Science and Technology, having been forced to form the consultation committee by social pressure, imperceptibly welcomed the opposition and did not adopt the proposal as an official stance of the ministry (Kim, 2005).

The NGOs, which had placed their hope in the governmental legal actions, were disappointed to see the government react rather passively, and requested passage of a bioethics law. They pointed to the fact that reproductive medicine was widely unregulated in South Korea, with more than 100 operating infertility clinics and virtually no restrictions in place on IVF treatment and surplus embryos. The Ministry of Health and Welfare released a proposal for a bill in a public hearing held on 6 December. The Ministry of Science and Technology, Dr Park Se Phil and other embryo cloning scientists strongly resisted such measures. In addition, Professor Hwang said, 'If this bill is approved, development of related world class technology would be badly hit.' In the face of such resistance, the Ministry of Health and Science announced that the bill was a mere research report.

In July 2001 civic groups held a press conference to urge passage of a bioethics law and embarked on the Campaign Group to Urge Rapid Enactment of Bioethics Law. The campaign body initially consisted of representatives from 33 NGOs, but the number of participants almost doubled to 69, as more organizations representative of Catholics, Protestants, Buddhists, environmentalists, feminists and medical associations took part in the campaign. Ever since, the campaign body has devoted itself to numerous activities to urge the enactment of a bioethics law through gatherings, announcement of joint declarations, submission of opinions, press conferences, discussion forums, a signature-seeking campaign, distribution of public relations articles and launching of websites. Clearly, hESC research, in a context of a political-institutional vacuum in the field of 'boundaries of life' topics, had become a highly controversial topic, with even government ministries taking the lead in preparing legislation providing a ban not only on SCNT research, but also on hESC research in general.

When in 2002, Clonaid – where, according to its founder Rael, some Koreans were involved – announced that a cloned human being would be born at the end of the year, the pressure to act in the field of cloning regulation intensified. The Ministry of Health and Welfare held the second public hearing in July to discuss a Life Science and Heath Safety and Bioethics bill. Prompted by the movement of Ministry of Health and Welfare, the Ministry of Science and Technology also prepared a bill to ban human cloning and research on stem cells. This proposed bill allowed embryonic cloning, which ignored the conclusions made by the Bioethics Advisory Committee formed under Ministry of Science and Technology. As conflict between ministries deepened and widened, giving rise to public criticism, the Office for Government Policy

Coordination ordered that the Ministry of Health and Welfare be the supervising ministry and consult with the Ministry of Science and Technology.

Until 2003 an open, deliberative process on hESC and cloning research operated in South Korea. But this situation changed dramatically when central policy actors such as the Roe adminstration imposed a new policy scenography destined to operate in the service of Hwang Woo-Suk, and Korean global leadership in hESC and cloning research. On 6 February 2003, a unified proposal for a bioethics law was announced by the government, and on 29 December in the same year, the proposal was passed by the National Assembly, which became effective in the year 2005. Research on human embryonic cloning, which had been the centre of debate, was legally authorized. Such policies ran counter to the original proposal made by Ministry of Health and Welfare and were also different from the report made by the Consensus Conference on cloning or the bioethics advisory committee under the Ministry of Science and Technology. Instead, the opinions of the Ministry of Science and Technology and the Ministry of Ministry of Commerce, Industry and Energy, both of which are in charge of fostering biotechnology, were reflected in full scope (Kim, 2005).

Throughout the political discussion of the new law, Hwang Woo-Suk was able to continue his research on SCNT. In 2005, when the bioethics bill came into force Hwang registered his research team as the SCNT Institute on January 3, and was authorized immediately. Also, on January 25, he was granted an authorization by the Minister of Health and Welfare to continue SCNT research. The law contained a clause stating that those who had been making progress in their research on SCNT 3 years before 1 January 2005, the date when the law was enforced, and those who had previously published their theses could continue their studies upon authorization by the Minister of Health and Welfare. The one and only person who has been granted an authorization on his or her research on SCNT according to the descriptions of this clause was Hwang (Han, 2006).

Despite these efforts of the Korean government not to 'put any obstacles into the way of Prof. Hwang', in May 2004 allegations began to surface of ethical misconduct in Hwang's lab, in particular, of Hwang having used oocytes for research donated from his laboratory workers. The accusations never stopped and eventually turned into one of the largest biomedical fraud and ethical misconduct cases in medical history (Gottweis and Triendl, 2006). As we saw in Chapter 2, Hwang had reported using 185 oocytes from 18 women for the first *Science* paper and 242 oocytes from 18 women for the second (Hwang et al., 2004b; Hwang et al., 2005). After his fall, a report of the National Bioethics Committee stated that from 28 November 2002 to 24 December 2005, at four medical institutions 2221 eggs were collected from 119 women with monetary compensations having been paid to 66 of them (National Bioethics Committee (NBC), 2006). It should be pointed out that the Hwang scandal did not originate in accusations of scientific misconduct but as a case of violation of medical ethics. The South Korean government's approach to support Hwang at all costs never reflected a social consensus and provides evidence of how quickly disastrous consequences could result from filtering out voices of critique and resistance in hESC and cloning research.

Conclusions

The regulation of hESC research and cloning is, first of all, characterized by radical uncertainty, ambivalence and an overflow of meaning in its core terminology and semantic architecture. In particular, the boundaries between 'cells', 'the embryos' and 'human beings' and their various associations such as rights, dignity and status have been in a constant state of flux in the countries discussed in this book. The stabilization and separation of the discursive field of hESC research from other areas in the field of life have been accomplished with varying success in the countries analysed.

First, as our comparison indicates, there seems to be a relationship between the *early, proactive and coherent effort to deal with new challenges in stem cell governance* as opposed to half-hearted, delayed and contradictory approaches. Compare, for example, the United Kingdom and Italy. At first glance, we are dealing with strict regulations in a Catholic country and liberal regulations in a Protestant country, something that seems to reflect strongly religiously coded differences in the definition of human life. But the story is much more complicated. Whereas in the United Kingdom the strongly dislocatory event of the birth of Dolly the lamb triggered a unified government response leading to new hESC regulations, in Italy the political response to Dolly lacked a coherent approach, was delayed and did not translate into a coherent policy approach. Combined with a general regulatory vacuum in the field of reproductive medicine, Italy turned during the 1990s into a 'Wild West' of reproductive medicine (at least in terms of the regulation of regenerative medicine), with strange professors such as the infamous Severino Antinori creating the impression of a 'world out of control'. This constellation played an important role in the rise of groups utterly opposed to hESC research and polarized Italy into two bitterly divided publics. This constellation was similar to the situation in the United States, where a regulatory vacuum in the field of regenerative and reproductive medicine played a certain role in partially radicalizing the various stakeholders in the energy field. A clear tendency in US politics to develop a coherent and proactive approach towards hESC research from Clinton to Bush was bound to fail against the background of a glaring regulatory vacuum in the field of reproductive medicine, and the absence of trusted institutions dealing with hESC research.

Second, the idea of a coherent, proactive approach towards life science governance should not be confused with modernist, hierarchical top-down political decision-making. The case of stem cell politics is not especially characterized by the adoption of novel, participatory decision-making mechanisms, neither during the political decision-making process, nor as a reaction afterwards. But it does demonstrate well the importance of *creating trust through a variety of policy scenographies involving different discursive and institutional mechanisms, designs and strategies.* Here, interactions with the various publics form an important element. If we look at the United Kingdom and Japan on the one side, and Italy and Korea since 2000 on the other side, we see two contrasting styles in interacting with the variously defined and self-defined publics of stakeholders. Partially shaped by historical pattern, in the United Kingdom a consultative and deliberative style of communicating through position papers, White Papers and open public consultations was characteristic for the policymaking process. Even in Japan, a country not known for its deliberative policy style, the highly sensitive field of hESC research gave rise to such policy interactive strategies. By contrast, neither in Italy nor in Korea after 2002 could such an approach be developed, particularly in Korea, with disastrous consequences for stem cell research and biotechnology industry.

Third, the creation of trust goes well beyond an engagement and shaping of publics. At the same time, with the HFEA a 'trusted institution', an institution with 'ethos', had been designated as the key institutional actor in hESC regulation in Britain. Whereas in the United States, Germany and Italy new bioethics institutions were created partially with the idea to face up to the new governance challenge of stem cell research, in the United Kingdom, a well-established institution quickly became a strong asset in securing trust for the emerging framework of regulation. In the United Kingdom, the HFEA was neither known for its bioethical expertise nor for its transparency, but instead for its

pragmatic and successful way of dealing with complex matters in reproductive medicine. The early involvement of the HFEA in shaping UK
stem cell governance was not only a result of path-dependence, but the
result of a deliberate and purposeful approach to create a system of regulation that combines competence with success and trust building. But the
aspect of trust and 'ethos' have not only become part of stem cell governance in the form of creating acceptance for stem cell research. As the
case of Germany, in particular, demonstrates, the 'ethos' of certain
trusted voices, such as of the German President Rau, against the context
of uncertainty can play a decisive role in tipping the balance of decision-
making to one or the other direction. Rau's outspoken rejection of stem
cell research gained so much weight in the German context because it
was not only a trusted voice in a constellation of deep insecurity but
also filled an institutional void where institutions such as HFEA in the
United Kingdom simply did not exist.

Fourth, the importance of proactive governance, trust and ethos in
the policymaking process, all point in our view towards the key role of
the setting or staging of the policymaking process in the field of life
sciences. Who gets the right to speak to which audiences is a question
not simply decided by constitutions and general characteristics of
political systems but is often decided ad hoc in particular policy constellations. Stem cell governance today operates under general conditions
of radical uncertainty and requires the simultaneous mobilization of
different publics, the creation of institutional spaces for articulating
emotions, concerns and anxieties, and the shaping of narratives that
create fixations when boundaries are fluid and architectures of meaning
are fragile. Participation does not always and necessarily offer the
answer to such constellations, but, as the UK example shows, can be an
important element in life science governance. While participation,
deliberation, transparency and, in general, linking up with the citizenry
seem to be an important aspect of contemporary stem cell governance,
facing up to the stem cell governance challenge requires a more complex intervention. Both the success of HFEA in the United Kingdom
and the partial failure of ad hoc created bioethics boards elsewhere
vividly demonstrate the importance and the pitfalls of trust-building
strategies in life science governance. While discursive memory, cultural
traditions and elements of path-dependency seem to deeply question
any chance of a non-nationally-based strategy of stem cell governance,
the importance of policy dramaturgy and the staging of proactive governance arrangements open up scenarios for transnational, global strategies of governing stem cell science.

6
Bioethics and the Global Moral Economy of HESC Science

Introduction

As the last chapter indicated, ethical reasoning and the creation of national bioethics committees or boards have played an important part in dealing politically with the regulatory challenges of hESC research. For the most part, these bioethical actors have been supportive of hESC research, and they have also played a central role in separating reproductive from therapeutic cloning. Such separations have been essential for the further development of regulations concerning hESC research and cloning for medical purposes. Although these bioethics institutions and argumentations have not necessarily reconciled existing conflict constellations, they have provided the vehicle whereby ethical reasoning could contribute to the solutions of often apparently irreconcilable conflicts. This is a critical contribution because the business of politics normally advances through manoeuvring, negotiation and the search for compromise. Standoffs may occur, but in the absence of a resort to force, agreements are reached through an adjustment of positions influenced by the relative power of the participants. It is sometimes thought that because cultural values are frequently deontological (absolute statements of right and wrong) they are therefore non-negotiable, particularly when they concern a cultural object as fundamental as the basis of human life. However, this view underestimates the inventiveness of the political process when substantial resources such as national economic advantage are at stake.

In this chapter we show how the political need to reconcile the promise of new health technologies with the cultural costs of scientific advance has been met by the evolution of bioethics as a political community, transnational network and bureaucratic device. Bioethics has become the political means for the creation of a global moral economy in which the trading and exchange of values is normalized and legitimated. This is a process that we can observe not only in the field of hESC research but also in many other areas of regenerative medicine, such as in tissue transplantation, or in others fields of biomedical research, such as in biobanking (Gottweis and Petersen, 2008). Bioethics is the apparently neutral currency with which cultural values can be measured, positions priced and deals arranged. Without an active moral economy, the progress of the economy of hESC science and industry would be severely constrained. In this respect, we will ask if bioethics might also be a means for allowing an assessment of hESC research to be made that is beyond the constraints of national narrativizations of hESC research and its implications. Many actors in constellations of conflict in biomedical research seem to envision bioethics as a 'new language and discourse' that will somehow overcome locally and culturally determined constellations of dissent and conflict.

We will ask, what are its political functions and why do national and international authorities view these functions as necessary and valuable in their dealings with cultural issues? To what extent has the performance of these functions promoted the emergence of a global moral economy underpinned by the neutral currency of bioethics? How does the moral economy actually work and how does its operation relate to the formation and legitimation of national and international policy? What are the structures and networks of global bioethics that support the functioning of the moral economy and to what extent can bioethics be seen as a coherent epistemic community? Finally, given this context, what are the main characteristics of the global bioethical discourse regarding hESC science, the balance of power within it and the regulatory policies that it legitimizes?

Bioethics and the global moral economy

In general, the political need for bioethics as the currency of a global moral economy arose because medical science was no longer able to research, develop and apply new health technologies while at the same time deal with cultural opposition and retain public trust in these technologies through the simple application of its scientific authority. Science could no longer legitimize science. It had always had 'medical ethics' as a legitimizing prop, but by the 1960s this had become, according to one observer, 'a mixture of religion, whimsy, exhortation, legal precedents, various traditions, philosophies of life, miscellaneous moral rules, and epithets' (Clouser, 1993, p. S10): politically unhelpful if not positively dysfunctional. Given the social and economic importance of medical advances, their potential for cultural conflict and the inadequacies of medical ethics, a mechanism was required that could engage with the science, reassure the public and assume visible bureaucratic form if necessary.

As coined by the American doyen of medical ethics Robert Veatch, 'bioethics' was intended to indicate a medical ethics that was 'not the same as the past ethics of physicians' (Veatch, 1991, p. 1; see also Reich, 1995). Reinforced by accounts such as Rothman's *Strangers at the Bedside*, a division was driven between the old, discredited ethics and the vision of a new bioethics untainted by the limitations of the past (Rothman, 1991). For the vision to have substance and be capable of bridging the gulf between discussion and action, it needed an epistemic identity that would engage readily with the requirements of regulatory bureaucracy. In the United States, in particular, globally the most influential though not the only source of bioethics, 'principlism' provided an important initial platform for the creation of that identity and credibility.

The task of principlism, as Albert Jonsen puts it, is to create 'the common coin of moral discourse' in order, one might add, to help resolve the cultural tensions created by medical scientific advance (Jonsen, 1998, p. 333). In this sense, principlism acts as the vehicle for science. Through the enunciation and application of a set of principles, standardized rules are established that enable the translation of different moral positions to a common metric capable of facilitating, usually on a cost-benefit basis, choices and decisions. For the principles to operate efficiently, they must combine to produce a system capable of commensuration (the discarding of information), predictability and calculability: the characteristics of a currency (Evans, 2000; see also Clouser and Gert, 1990). So the principles themselves have embedded utility criteria that effectively rule some types of moral argument in and others out. There can be little doubt that the emergence of principlism was driven by government as a result of the recognition of the political need for its existence. The production of *The Belmont Report* in 1978 by the National Commission for the Protection of Human Subjects of Biomedical and Behavioral Research was closely linked to the work of members of the Kennedy Institute and their subsequent publication of *Principles of Biomedical Ethics* (National Commission for the Protection of Human Subjects of Biomedical and Behavioral Research, 1978; Beauchamp and Childress, 1989). Bioethics emerged because it was politically useful, and as Jonsen observes, principlism 'met the need of public policy makers for a clear and simple statement of the ethical basis for regulation of research' (Jonsen, 1994, p. xvi, quoted by Evans, 2000, p. 34). It was transparent, formally rational and bureaucratically friendly, value-free and apparently impartial (Lopez, 2004, p. 889). Given these qualities, principlism's political rise was rapid and was consolidated when it became the legally required decision-making system for recipients of US federal research funds.

As with all political movements, its rise was not uncontested – but it was consistent. In the early days of the human genetic engineering debate, theologians were clear that there should be a 'thick debate' about ends as well as means that included the public. This was opposed by scientists who 'pushed for a shift from public control over the ethics of human engineering through legislative activity to bureaucratic control through government advisory commissions' (Evans, 2002, p. 6). With bioethicists willing and able to service such committees, a style of ethical debate emerged that existed primarily within a bureaucratic context: formally rational, efficient and undisturbed by the volatilities of public emotion. However, even at the time there were arguments about whether this strategy was appropriate (DuBose et al., 1994). To an extent, the arguments over the soul of bioethics continue today, with questions regarding the relationship between bioethics, biopolitics and ideology, but not with sufficient weight to disrupt its policy utility (Callahan, 2006; Bishop and Jotterand, 2006; Koch, 2006).

By establishing itself as the state-sanctioned authority for converting discussions of good and bad in American medical science into a common language and concepts, the bioethics of principlism achieved the status of an ascendant political currency with global potential. If the qualities that had proved so attractive to the US government could be exported to countries and regions facing similar political problems in their handling of new health technologies, bioethics would be able to establish a dominant position in the international moral economy. But for this to occur, it would need ways of marginalizing competitors such as the Catholic Church while demonstrating its unique competence to government.

Given the pluralist base of many governments, it is important that bioethics presents itself and its methods as inclusive. Counternarratives are incorporated within the overall hegemony through the construction of 'legitimate differences' in the discourse of morality that are then subject to the application of bioethics principles (e.g., autonomy, beneficence, non-malfeasance and justice) (Laclau and Mouffe, 1985).

As a consequence, moral positions are translated into the common language of bioethics and are subject to the formal rationality of its discursive search for compromise, regardless of whether the proponents of these positions regard this as a sensible exercise (Evans, 2002, pp. 6–8). Since the whole point of the exercise is that a decision should emerge from it, outright opposition to the bioethical search for compromise is seen as a disqualification for which the penalty is marginalization, if not ostracization, from the regulatory process.

As the emerging currency of the international moral economy, bioethics presents itself as a neutral technique that is able to produce 'a single, correct solution for each ethical problem which is largely independent of person, place or time' (Bosk, 1999, p. 63). It is thus an eminently transferable coinage because its value is derived from its impartial functionality for the governance of science rather than from any localized source of historic or cultural authority such as religion. This degree of global utility helps explain the robustness of principlism in the face of challenges to its domination of bioethics from other philosophical approaches such as casuistry, narrative ethics and pragmatics. These have been readily repelled with the result that over time there has been little change 'in the contours, context, style of thought or the ideology of bioethics' (Fox, 1999, p. 11). Instead, the grip of principlism on bioethics has tightened, its self-confidence has grown and it has taken to asserting explicitly its championship of 'universal ethical principles' (Hedgecoe, 2004, p. 125).

But its growing hegemony over the governance of new health technologies in the global moral economy is also due to its adaptive qualities. Although its genesis was associated with the institutions of medical education and medical ethics, it rapidly acquired an enhanced legitimacy through alliances with other knowledge territories such as law, economics and moral and political philosophy, thus broadening the sources of its epistemic power (Lopez, 2004, p. 888). At the same time, its political flexibility is demonstrated in its ability to evolve fresh forms of principlism in order to maintain its governance lead as the arbiter of cultural tensions. Thus, in 1994 Knoppers and Chadwick identified five basic principles underlying what they termed the 'international consensus' on research on the human genome: autonomy, privacy, justice, equity and quality out of respect for human dignity (1994). Nine years later, they concluded that, as a result of new genetic research, these existing principles had been superseded by new international trends in ethics that should be collated and codified as the principles of reciprocity, mutuality, solidarity, citizenry and universality (Knoppers and Chadwick, 2005). From the perspective of Knoppers and Chadwick as leading bioethicists, the ad hoc evolution of bioethics through the adaptation of its initial principles to local technological development at the national level was acceptable but needed order imposed upon it.

For just as national and religious cultures vary in their response to advances in medical science, so also do national forms of bioethics. In her study of science in the United Kingdom, United States and Germany, Jasanoff notes the dawning recognition in all three political systems that some of the risks and promises engendered by the multi-faceted advances in genetics and biotechnology

> called for a new language of deliberation, *geared to the analysis of human values* rather than the benefits of the market, the facts of science, or the norms of law. The language that actors seized on for this purpose was a branch of moral philosophy, specifically, bioethics. Combining life (bios) and moral custom (ethos) in a single portmanteau word, bioethics offered the promise of bringing order and principle to domains previously governed by irrational, emotive and unanalysed reactions.
>
> (Jasanoff, 2005b, p. 171; emphasis added)

The construction of bioethics took different paths in the three countries, each influenced by their national philosophical traditions and, in the German case, by the continuing search for national identity. The moral and political concerns were also different. In the United Kingdom the driver was the need to safeguard a space for science, in the United States the minimization of the risk to patients and in Germany the protection of human dignity. Nonetheless despite the contrasting moral foci of the three countries, 'bioethics was seen in all three countries as a device for bridging potentially troublesome divides: among disciplines, professions, and institutions; and increasingly also among science, state and society' (Jasanoff, 2005b, p. 188). There was a common utilitarian emphasis on the identification of ethical procedures that could be used to address and perhaps resolve conflicting moral positions in ways that may be integrated with the policy process.

Bioethics is thus able to position itself in the political space situated between the cultural tensions created by advances in medical science and the public trust on which that advance depends. Precisely how this 'political positioning' is manifested in an institutional location for bioethics within, or adjacent to, the policy process will vary from country to country in response to national regulatory cultures and ethics strategies: an institutional experience that is shared with technocratic forms of governance (Rothstein et al., 1999). Thus in the United Kingdom, the Nuffield Council on Bioethics is an independent body at arms length from government, whereas in Germany the National Ethics Council is a government-appointed body with the brief to act 'as a national forum for dialogue on ethical issues in the life sciences'. But both are acting as forms of what Jasanoff terms 'official bioethics' and Kelly as 'public bioethics' in their promotion of societal consensus through a language that claims to be morally neutral (Jasanoff, 2005b, pp. 173–88; Kelly, 2003). In terms of their institutional relationship with policy formation on the regulation of science, bioethics committees perform a similar political function as those of the scientific advisory system: agenda setting on the basis of an expert authority that can be used by government to legitimize the subsequent regulatory policy outputs (Salter and Jones, 2005, pp. 712–15). The difference is that whereas the authority of the scientific advisory system is technocratic, that of bioethics committees is ethical. Once formed and implemented, the regulatory policies then become a political product at least partly shaped by the bioethical agenda.

Global ambition and networks

The transition of bioethics from a set of ideas produced to deal with local national problems to a global movement confident that it can be applied to any of the cultural issues raised by the advances of medical science has been rapid. Both the movement's self-belief and its awareness of its political potential are palpable. The European Commission's European Group on Ethics in Science and New Technologies (EGE), for example, has no doubts about its role as a political broker and sees its opinions produced for the Commission as 'a necessary stage in the debate about the relationship between new scientific breakthroughs and the evolution in attitudes that this progress brings' (European Group on Ethics in Science and New Technologies, 2001, p. 2). Equally trenchant are Knoppers and Chadwick when they argue for the codification of their principles in an international instrument that national governments would apply through legislation, review and oversight. They fear without the guiding hand of bioethics that not only the accountability of the HGP (with which they are primarily concerned) is at stake but 'so also are our present obligations of stewardship to human kind and to future generations. This unique opportunity to provide *principled direction* must not be lost' (Knoppers and Chadwick, 1994, p. 2036, emphasis added). UNESCO's International Bioethics Committee (IBC) clearly agrees with them, and in June 2005 produced its *Universal Draft Declaration on Bioethics and Human Rights* in pursuit of its ambition of an authoritative statement of and guide to global bioethics values. Yesley notes the growing political maturity of bioethics, observing that the January 2005 meeting of the IBC where the upcoming bioethics declaration was discussed 'was an opportunity to witness several ongoing transitions in bioethics: from a philosophical to a legal orientation, from national to international standards, and from professional to political policy making' (Yesley, 2005, p. 8).

Inspired by an apparent sense of moral duty, the rise of bioethics as a political force in the global moral economy – particularly in areas of medical science that deal directly or indirectly with the human embryo, the human genome and human identity – has been swift. Its ascent has been marked by a global infrastructure: high profile ethical statements launched from the platform of established international bodies, an awareness of the need to translate these statements into legal/ bureaucratic form, and the proliferation of horizontal and vertical networks linking international and national levels of ethical governance.

The political impetus for the internationalization of bioethics was established on 11 November 1997 when the General Conference of UNESCO adopted the *Universal Declaration on the Human Genome and Human Rights* (Declaration), which, in its own words, forms part of 'a framework of thinking, known as bioethics, which relates to the principles that *must guide* human action in the face of the challenges raised by biology and genetics' (UNESCO, 1997; emphasis added). Behind the Declaration lay the work of UNESCO's International Bioethics Committee, a body established in 1993 and composed of 36 independent experts. Beside the IBC sits the Intergovernmental Bioethics Committee (IGBC – created in 1998), composed of representatives from 36 Member States of the UN, that translates the IBC's advice into its own proposals and recommendations to the Director General for transmission to Member States, the Executive Board and the General Conference. The IGBC thus serves to give the IBC both representative legitimacy and a solid bureaucratic position in the UN's policymaking machinery.

Propelled by the continuing activities and reports of the IBC, the Declaration acted as the template and rationale for a series of bioethical initiatives by other international organizations such as the Council for International Organizations of Medical Sciences (CIOMS), the Human Genome Organisation (HUGO), the International Society of Bioethics (ISB), the World Medical Association (WMA) and the World

Health Organization (WHO) (see, e.g., CIOMS, 2002; HUGO, 1996; ISB, 2000; WMA, 2002; WHO, 1997). Once the momentum was established, it rapidly became unacceptable for an international organization concerned with medical science not to have at least a reflective, if not a promotional, ethical function somewhere within its structure. Thus the WHO, for example, in October 2002 launched its Ethics and Health Department 'to provide a focal point for the examination of the ethical issues raised by activities throughout the organisation' (WHO, 2005). The department's functions include a global calendar of bioethics events, resources on research ethics and support for the annual Global Summit of Bioethics Commissions. Working through the WHO's six regional offices and what it terms 'regional ethics focal points', the department provides linkage between the international and national levels of bioethics and facilitates the development of the emerging transnational bioethics network.

At the regional level of the EU, we find that the principles of the IBC's *Universal Declaration on the Human Genome and Human Rights* are faithfully reflected in a similar list of rights incorporated in the Council of Europe's 1997 'Convention on Human Rights and Biomedicine' (popularly known as the 'Bioethics Convention' and drawn up by the Council's Steering Committee on Bioethics (CDBI)). Significantly, the convention is legally binding on countries that sign and ratify it, and each state that does so is obliged to bring its laws into line with it. The Bioethics Convention also gained the support of the European Parliament (European Parliament, 1997). Having established its international presence through the continuing influence of the Convention, bioethics gained substantial institutional legitimacy in Europe through the activities of the EGE. Accountable to the European Commission, the EGE has acquired bureaucratic respectability through its bioethical contribution to the resolution of practical decision-making problems in the EU's dealings with human genetics (Salter, 2006b). At critical points in the life of FPs 5 and 6 (the EU's research funding programmes), the EGE has been summoned by the European Commission to act as a respected political broker over difficult regulatory issues relating to human embryo research and human genetic technologies (Salter and Jones, 2002, pp. 812–3).

However, although there is much evidence to confirm the rapid political rise of bioethics in terms of an infrastructure of transnational networks, the dissemination of ideas and institutional engagement with policymaking at both national and international levels, other data also suggest that it is yet some distance from becoming what one might term a hegemonic epistemic community in control of a particular knowledge terrain (Haas, 2001).

Although principlism may be a dominant force within the philosophical community of bioethics, and although its language and precepts may inform the reports of national and international bioethics committees, philosophers are not necessarily dominant in the membership of these committees. Rather, it is noticeable that members of such bioethics committees regard it as appropriate to describe themselves by their established, non-bioethics, disciplinary backgrounds. Indeed, multidisciplinarity is regarded as a strength, and the absence of members specifically trained in bioethics is clearly not regarded as a weakness (Galloux et al., 2002). In other words, the business of bioethical regulation may require a particular expertise, but it is one which, it is believed, can be acquired through serving an apprenticeship in the practical trade of committee work and ethical problem solving rather than through acquiring the formal ethical knowledge of philosophical training. Hence, a study of the bioethics committees of national biobanks, the IBC and EGE found that out of a total membership of 88 only 11 (12 per cent) describe themselves as medical ethicists, philosophers or theologians. The dominating characteristics of membership background were medical science, particularly medical genetics, and law with 55 members (62 per cent) thus described (Salter and Jones, 2005). Given this mix of backgrounds in the bureaucratic identity of bioethics, it may be more appropriate to describe the political community of bioethics not as a single epistemic community but, as Rosenberg observes, as a bioethical enterprise comprising 'a conglomerate of experts, practices, and ritualised and critical discourses in both academic and public space' (Rosenberg, 1999, p. 40). It is the political utility of this enterprise and the common adherence of its members to principles and procedures that can produce a usable decision that constitutes its principal identity and strength within the global moral economy.

Bioethics and hESC science

The inevitable engagement between hESC science and the moral status of the human embryo creates the potential political need for bioethics to act as a vehicle for the resolution of cultural conflict. At the same time, the international movement of hESC science and its accompanying economic potential has rendered it a global phenomenon, which, given the parallel development of bioethics with its interest in the facilitation of medical advance, is likely to result in a global policy discourse characterized by value trading in the world's moral economy.

In their study of the stem cell discourse in the United States, Wolpe and McGee note that in public policy debates 'the first battle is often a struggle about definitions, and the winning side is usually the one most able to capture rhetorical primacy by having its definitions of the situation accepted as the taken-for-granted landscape on which the rest of the game must be staged' (Wolpe and McGee, 2001, p. 185) – that is, the debates established the agenda within which policy was to be formed. When viewed in the global context, the formal bioethical discourse initially defined a common language with a limited set of components consisting of the building blocks of hESC science. These are thus simultaneously scientific and ethical objects (Table 6.1).

Through the use of ethical arguments for or against the use of particular components, or combinations of components, hierarchies of meaning are created and choices implied about the preferred future path of hESC science. Given the scientific, economic and cultural pressures at work in the field, positions over time are rarely static and cultural trading occurs in the sense that a government or authority may, in response to these pressures, alter its support for particular components, or set of components, in order to achieve political advantage or a workable political compromise. In addition, new scientific possibilities combine with the inventiveness of bioethicists to generate new cultural components to the discourse.

In very rough fashion, the political significance of hESC science can be gauged by the number of national and international bioethical committees that have deemed it appropriate, or have been requested by their governments, to produce a report on the regulation of the science (bearing in mind the early development of hESC science and the

Table 6.1   The units of cultural trading

- Embryo source
  - aborted
  - IVF supernumerary
  - non-IVF donated
  - cloned
- Embryo creation date
- Embryo age
- hESC line origin
- hESC line creation date
- hESC line research purpose

wide range of other new and potential health technologies demanding bioethical attention).

In the EU, a 2004 survey of the national bioethics committees (or equivalent bodies) of the 25 Member States provides initial insights into the political salience of hESC science and the contribution of the committees to the discourse on its regulation (Table 6.2). Of the 25 countries, two-thirds (16 countries) had considered hESC science sufficiently important to require an opinion from their national ethics committee (or similar body) on the ethical and policy issues involved. In addition, two-thirds had initiated, or intended to initiate, a public debate on the new technology that frequently engaged the wider national bioethics communities. Finally, half of the countries (13) had engaged in both expert and public discussion. So far as the majority of the EU Member States are concerned, as a governance issue deserving serious bioethical consideration, hESC science is clearly politicized. Given this evidence, it is not surprising that the EU institutions themselves (Commission, Council of Ministers and European Parliament) have been similarly exercised by bioethical debate over the conditions under which hESC research should, or should not, be funded.

Contributions to the global debate have also come from the national bioethics committees of other countries with an interest in the development of hESC science such as China, India, Israel, Singapore and the United States (China Ministry of Science and Technology and Ministry of Health, 2004; Indian Council of Medical Research, 2004; Israel Academy of Sciences and Humanities Bioethics Advisory Committee, 2001; Singapore Bioethics Advisory Committee, 2002; US President's Council on Bioethics, 2002, 2004, 2005). And at the international level, both UNESCO's IBC and the Human Genome Organisation's Ethics Committee have produced position statements on hESC science (HUGO, 2004; UNESCO, 2001). If you are, or want to become, a major player in hESC politics, then having a formally approved bioethical position that legitimates your policy stance is now a requirement of participation in the game.

The activation of the global moral economy in the field of hESC science by numerous bioethics committees and the exploration of the various possible configurations of its component cultural units have resulted in five major ethical trading positions. The options are

1. Prohibition of derivation of hESCs from human embryos
2. Prohibition of derivation of hESCs but allowing importation

Table 6.2 National Bioethics Committees and hESC science (EU Member States)

Country	National ethics committee opinion on hESC science	Public debate on hESC science
Austria	Yes – Die Bioethikkommission	Yes
Belgium	Yes – Comité Consultatif de Bioéthique del Belgique	No
Cyprus	No – Cyprus National Bioethics Committee	Yes (intended)
Czech Republic	No	Yes
Denmark	Yes – Danish Council of Ethics	Yes
Estonia	No	No
Finland	No	No
France	Yes – Comité Consultatif National d'Ethique	Yes
Germany	Yes – National Ethics Council	Yes
Greece	Yes – National Bioethics Commission	No
Hungary	No	No
Ireland	Yes – Commission on Assisted Reproduction	Yes (intended) – Irish Council on Bioethics
Italy	Yes – Italian National Bioethics Committee	Yes
Latvia	Yes – Central Medical Ethics Committee	Yes
Lithuania	No – Lithuanian Bioethics Committee	Yes
Luxembourg	Yes	Yes
Malta	No	No
Netherlands	Yes – Health Council	Yes
Poland	No	No
Portugal	Yes	No
Slovak Republic	No	No
Slovenia	Yes – National Medical Ethics Committee	Yes
Spain	Yes – Advisory Committee on Ethics of Scientific and Technical Research	Yes
Sweden	Yes – Swedish National Council on Medical Ethics	Yes
United Kingdom	No single national committee, although House of Lords Select Committee has played a key role	Yes

Source: European Commission, DG Research (2004), *Survey on Opinions from National Ethics Committees or Similar Bodies, Public Debate and National Legislation in Relation to Human Embryonic Stem Cell Research and Use*, volume I in EU Member States (Brussels: European Commission).

3. Allowing derivation of hESCs from supernumerary human embryos
4. Prohibition of creation of human embryos for research purposes including cloning
5. Allowing the creation of human embryos for research purposes including cloning

In cultural terms, the bioethical options are constructed so that as one moves from 1 to 5 the moral status (value) attached to the human embryo diminishes and the value attached to its scientific, commercial and social utility correspondingly increases. In regulatory policy terms, the same continuum can be described as extending from 'restrictive' (of scientific freedom through protection of the embryo) to 'liberal' (facilitative of science, industry and certain social interests) – though these terms, of course, have their own value connotations. In political terms, the continuum shows the increasing impact of scientific, industrial and pro-hESC social interests as one moves from Option 1 to Option 5.

The political utility of the moral economy of bioethics is that its products can be, and are intended to be, translated into policy options in order that in an ideal bioethical world the policy is perfectly legitimated by the ethics. If we apply the same categories to the realm of national policy, we find a global distribution of regulatory policies summarized in Table 6.3. Formal government policy is indicated by the presence of an 'X' and the absence of a policy by an empty cell. Within the five categories, states have adopted individual variations in response to their local cultural pressures.

Of the 39 countries on which data was available, only 7 have adopted Policy Option 1, the prohibition of the derivation of hESCs from human embryos. While the transnational cultural influence of Roman Catholicism is undoubtedly present in the examples of Austria, Ireland, Italy and Poland (although Spain has moved towards liberalization), Germany presents an interesting case in which a range of cultural factors have converged. As we saw in Chapter 4, the experience of the Third Reich and of Nazi eugenics has joined the churches, the Green Party and numerous social movements in an alliance against hESC research. Yet despite substantial cultural opposition, science has still achieved a political compromise using the ethical object of 'hESC line origin'. Both Italy and Germany have adopted Policy Option 2 in which prohibition of the derivation of hESCs from human embryos is retained but the importation of hESC lines is allowed. In addition, Germany has imposed the extra conditions of 'hESC line creation date' and 'source of embryo': Its Stem Cell Act 2002 requires that the hESCs

Table 6.3 The global regulation of hESC science (2004)

Countries	Policies				
	Option 1 Prohibition of procurement of hESCs from human embryos	Option 2 Prohibition of procurement but allowing importation	Option 3 Allowing procurement of hESCs from supernumerary human embryos	Option 4 Prohibition of creation of human embryos for research purposes, including cloning	Option 5 Allowing creation of human embryos for research purposes including, cloning
Australia			X	X	
Austria	X			X	
Belgium					X
Canada			X	X	
China					
Cyprus				X	
Czech Rep.				X	
Denmark			X	X	
Estonia			X	X	
Germany	X	X		X	
Greece			X	X	
Finland			X	X	
France			X	X	
Georgia				X	
Hungary			X	X	
Iceland				X	
India			X		X
Ireland	X			X	
Israel			X		X
Italy	X	X		X	
Japan			X		X
Lithuania	X			X	

(*Continued*)

Table 6.3 (Continued)

Countries	Policies				
	Option 1 Prohibition of procurement of hESCs from human embryos	Option 2 Prohibition of procurement but allowing importation	Option 3 Allowing procurement of hESCs from supernumerary human embryos	Option 4 Prohibition of creation of human embryos for research purposes, including cloning	Option 5 Allowing creation of human embryos for research purposes including, cloning
Luxembourg					
Latvia				X	
Malta				X	
Mexico				X	
Netherlands			X		
Norway					
Poland	X				
Portugal				X	
Singapore					X
South Korea			X		X
Spain			X	X	
Sweden			X		
Slovenia				X	
Slovakia	X			X	
Switzerland			X	X	
Taiwan			X	X	
UK			X		X
Total	7	2	19	26	7

Sources: European Commission, DG Research (2004), *Survey on Opinions from National Ethics Committees or Similar Bodies, Public Debate and National Legislation in Relation to Human Embryonic Stem Cell Research and Use*, volume I, EU Member States; volume II, Countries associated to FP6 and third countries (Brussels: European Commission); Walters R (2004) 'Human Embryonic Stem Cell Research: An Intercultural Perspective', *Kennedy Institute of Ethics Journal*, 14(1), 3–38.

were derived from supernumerary embryos before 1 January 2002. Both hESC line origin (importation) and hESC line creation date are ethical objects that seek to place at least some moral distance between governments and the act of embryo destruction necessary for hESC creation. They are useful for manoeuvring in small political spaces, hence President Bush's use of an hESC line creation date when he announced his decision to allow US federal funds to be used for research on existing (that is, before 9 August 2001) hESC lines.

A further refinement and reduction in the moral value of the embryo is incorporated into Policy Option 3, allowing derivation of hESCs from supernumerary human embryos. This policy is constructed by viewing 'embryo source' as a set of possible choices and then arguing that the IVF supernumerary embryo is the morally superior option over the aborted, non-IVF donated and cloned embryos. Some embryos are more valuable than others. Within this option the low intrinsic moral value of the embryo is matched with its high utility value to the progress of science. In an interesting compromise used by some countries, the low moral value of the supernumerary embryo (and hence its lack of self-protection) can to an extent be mitigated by an 'embryo creation date' criterion. Under the terms of Australia's Research Involving Human Embryos Act 2002, for example, supernumerary embryos used in research must have been created before 5 April 2002 in order to qualify as what the Act terms 'excess' embryos (an interesting addition to the concept of 'spare' embryos). So the ethical criterion becomes 'when' the embryo is created as well as 'how'.

The political attractions of Policy Option 3 are practical as well as ethical. IVF is an established international industry, and its extension as a supplier for hESC science provides an unexpected commercial windfall for products that would otherwise have little or no market value. At the same time, governments can be seen to be taking both an ethical stand and to be facilitating the advancement of hESC science. If this happy combination is then linked to Policy Option 4, the prohibition of the creation of human embryos for research purposes including cloning, then the ethical legitimacy of the government is consolidated. Partly for this reason, perhaps, we find that 11 of the 18 countries pursuing Policy Option 3 have also implemented Policy Option 4.

Policy Option 4 is easily the most popular option with two-thirds of countries adopting it either through national law or through the ratification of the Council of Europe's *Convention on Human Rights and Biomedicine* (Article 18 states: 'The creation of human embryos for research purposes is prohibited'). In the global politics of hESC science, the latter has proved to be an instructive example of how international conventions can exercise a shaping influence on the development of policy domains. Policy Option 4 can be seen as the last barrier to the final dismantling of the moral status of the human embryo. Beyond it lies Policy Option 5, allowing the creation of human embryos for research purposes, including cloning, whereby the embryo has no separate moral status at all but is simply an instrument in the application of hESC science. Here the cultural values of scientific freedom, economic progress and the right of citizens to new medical therapies are paramount, and the issue of the moral status of the embryo is no longer an inhibiting factor. Only seven countries have yet achieved this particular ethical balance.

As the expression of the dynamic of cultural politics and the contested construction of ethical meaning, the global continuum of regulatory policies is not static but contains what appears to be a distinct trend for countries to move progressively from left to right from Policy Option 1 and towards Policy Option 5 (Italy is the sole example of national policymaking moving towards a more conservative position as a result of its 2004 Assisted Fertility Law). Over the last 5 years, countries that have introduced policies more facilitative of hESC research than policies that existed before include Australia, Denmark, France, Germany, Greece, Japan, the Netherlands, Singapore, South Korea, Spain, Switzerland and the United Kingdom. Apart from Italy, there are no examples of countries introducing new policies that are less facilitative of hESC science. Even in the seven states where Policy Option 1 prohibits the derivation of hESCs from human embryos, indications exist that the pressures for change may render this position unsustainable in the medium to longer term. Italy and Germany have already shown they are not immune from the global moral economy and its facilitation of the trading of cultural values by their acceptance of Option 2 (prohibition of derivation but allowing importation of hESCs). As the transnational networks of hESC science and industry continue to develop and the international competition increases, further ethical compromises may emerge. At the national level, the impression thus far is one of a lone Catholic Church fighting a rearguard action that is gradually being lost – which is one reason why the regional and international levels of governance acquire extra political significance.

However, although the majority of countries have progressed to the adoption of Policy Option 3, further movement to Policy Option 5 may prove more difficult. The global pace of cultural trading, value shift and policy change may now slow. The reasons for this are several. Most obviously, two-thirds of states with regulatory policies on hESC science have adopted Policy Option 4, the prohibition of the creation of human embryos for research purposes, including cloning, in many cases as a result of the signing and ratification of the Council of Europe's *Convention on Human Rights and Biomedicine*. Withdrawal from the Convention would be visible and risk political exposure. In addition, the political linkage between cloning for the purposes of reproduction and cloning for the purposes of scientific research into disease ('therapeutic cloning') that the Vatican, the United States and Costa Rica have sought to establish in the UN arena has complicated the process of policy choice (L. Walters, 2004). Through the alignment of the issues of (a) cloning and (b) deliberate embryo creation for research purposes, a broader range of cultural values is stimulated with a consequent expansion in the complexity of the cultural trading required to achieve a bioethical and policy solution. (This problem highlights the ethical advantages of Policy Option 3 with its condition of *spare* embryo use.)

Conclusions

The political utility of bioethics to the advance of medical science lies in its ability to normalize the trading of values between conflicting cultural positions, so an ethical solution is produced that can then be translated into legitimate regulatory policy. Its claim is to neutrality, impartiality and the application of rational principles that reduce different value positions to a common bioethical currency. In the main, that currency now has international credibility and forms the basis for a growing global moral economy in which values regarding the moral status of the human embryo can be traded as part of the search for political compromise. Cultural values may be contested, but the contest is routinized and assumed to be productive through the exchange of values.

The success of the bioethical enterprise lies in its ability to link 'bureaucracy – committees, institutional regulations, and finely tuned language – with claims to moral stature' (Rosenberg, 1999, p. 38). This quality has produced a burgeoning infrastructure of support for the bioethical currency in terms of national and international committees, networks and organizations dedicated to standardizing the main currency units of value while simultaneously encouraging their refinement into ethical subunits where this is deemed to be functional. To varying degrees, bioethics committees are becoming embedded within the policy process and asked to deal with the cultural issues raised by medical science.

But, as we have shown in the previous chapter, the creation of bioethics institutions as such did not necessarily correlate with an alleviation of political conflict over hESC research in different political systems. If created proactively, the impact of bioethics institutions tended to be much greater than when they were set up 'ad hoc' in order to settle and negotiate already existing conflict situations.

In the hESC field bioethics can enable the interrogation of ethical options, and thus the refinement of the political currency of the moral economy, through the investment of ethical significance in such characteristics as the source, date of creation, age and research purpose of the embryo or ESC. For the most part, the compromises they have produced and that have been embodied in regulatory policy have been supportive of hESC science – though, as part of the ethical trading, scientists have frequently had to accept various constraints on their activities. In a sense, this is not surprising. The data on the membership of national bioethics committees, at least, indicate that the majority are medical scientists or lawyers; neither group has a history of resistance to scientific progress. If bioethical philosophers are not well represented on such committees, they nonetheless have an important political contribution to make to the vitality of the global moral economy through the creation of fresh ideas and principles. The epistemic identity of bioethics may be fragmented, but this does not undermine its functionality.

In the course of carrying out their political task of reconciling cultural concerns with scientific progress, bioethics committees have become an increasingly visible political site as competing interests search for influence over their membership, agenda setting and discussions. For example, the 2005 election of five practising Roman Catholics among the nine new members to the European Group on Ethics provoked concern among some of the other EGE members, given the Catholic Church's position on human embryo research (Bosch, 2005). Nonetheless to date bioethics committees have, broadly speaking, been able to deal with the political pressures placed upon them in the field of hESC science, but this may not continue as public interest in its therapeutic applications develops. Alternatively, the balance of power in the ethics–science relationship may need to change. Already we find the ingenuity of bioethicists being stretched by the demands placed upon them, and as a response, their requests that science help them find a way out of ethical dilemmas. Thus the US President's Council on Bioethics (PCB), which has consistently placed a high value on the moral status of the human embryo (see PCB, 2002, 2004), recently sought a response from science to its suggestions for obtaining pluripotent, genetically stable, and long-lived human stem cells (the functional equivalent of hESCs) that do not involve creating, destroying or harming human embryos (PCB, 2005, p. x).

What may emerge from this kind of interaction is renegotiation of the units of bioethical currency so that rather than the operation of the global moral economy being guided purely by scientific categories of value, it becomes at least partially dependent on cultural categories of value. Such renegotiation would challenge the automatic instrumentality established by the dominance of principlism in the politics of bioethics. However, it is unlikely that such a challenge would result in anything other than a temporary slowing down in the progress of hESC science. Intra-bioethical conflict may become more common, the bioethical currency more refined and the possible range of ethical positions on hESC regulation more complex, but the fundamental dynamic of the global moral economy as a vehicle for the support of science, energized and facilitated by the struggle for national advantage, will not be diverted.

7
HESC Science and the Cultural Politics of the EU's Framework Programmes

Introduction

It is estimated that when all of the EU's Framework Programme (FP) 6 (2002–6) projects have come to an end, about €21 million will have been spent on hESC research. This figure constitutes 0.85 per cent of the €2.45 billion Health Research Programme budget within FP6 and 0.10 per cent of the total €17.5 billion budget (Europa, 2007). If the financial commitment of the EU to hESC science can be shown to be very small, the EU's moral commitment has nonetheless been very large. For at least a decade, debates have raged throughout EU institutions about the values that should inform the funding of this field. Indeed, such have been the divisions on the issue that they threatened to prevent the approval of FP5 (1998–2002), FP6 and its successor FP7 (2007–13).

The inverse relationship between the size of the hESC research budget and the political effort expended in gaining approval within the FP is a product of the intense cultural politics surrounding this new science in the several arenas of the EU. As we have seen in Chapters 4, 5 and 6, hESC research's reliance on the destruction of the blastocyst for the production of stem cells provokes a collision at national and international levels between the demands of science and cultural values that ascribe a high moral status to the early human embryo. The political negotiation of this collision is frequently difficult, but when viewed on a global scale, can be seen to be facilitated by the use of bioethics as a form of expert governance. Precisely how effective bioethics can be in this political role is dependent on both its authority and the nature of its linkage with institutional decision-making. States vary in their use of bioethics as a formal aid to government and in the extent of its integration with the state bureaucracy.

This chapter analyses the cultural conflict over the funding of hESC science in the FPs as reflected in the documents and debates of the European Commission, the Council of Ministers, the European Parliament and numerous committees. It shows how bioethics has facilitated the emergence of the 'moral economy' identified in Chapter 6 wherein values may be traded and cultural disputes routinized, though not necessarily resolved. We will also see how a bioethical institution can develop significant impact on the policy process through proactive manoeuvring and a strategy to generate trust rather than one that focuses simply on the quick facilitation of the political decision-making process.

The approach of the chapter is as follows. First, having outlined a particular understanding of cultural politics, it discusses the formal consolidation of the position of bioethics in the EU in terms of the rise of the EGE and its contribution to the resolution of the difficulties over hESC research in FP5. Second, it explores in detail the bioethical discourse of FP6, the institutional manoeuvring around whether hESC research should be funded and if so on what ethical conditions. Third, it examines the bureaucratization of bioethics and its incorporation as an integral part of decision-making on project applications for FP funding. Finally, the chapter reflects on how conclusive the recent FP7 debates over hESC science are in revealing a stabilization of the cultural conflict within particular dimensions.

Cultural conflict and the EGE

How can we best conceptualize the cultural politics generated by the demands of hESC science and the contribution of bioethics to the consequent competition for control of political meaning? Social anthropology's approach to the concept of 'culture' and 'cultural politics' has evolved in ways that render it a useful tool for understanding how values interact with the operation of power. Originally, anthropology treated 'culture' as if it were a set of ideas or meanings shared by a whole population of homogeneous individuals (Asad, 1973, 1979). 'Culture' thus became a bounded entity with a fixed identity that sustained itself in a static equilibrium (Gough, 1968). However, later approaches have challenged this interpretation and emphasized instead the dynamic nature of culture 'as an active process of meaning making and contestation over definition, including itself' (Street, 1993: 2). Clearly this approach resonates much more readily with the concept of 'politics' where the struggle for power is a given that is rarely, if ever, static. It makes it clear that culture is a resource that can form part of the political action in terms of both context and content.

Once 'culture' is seen as a contested process of meaning making, other questions about key terms and concepts arise. In her discussion of the politicization of culture, Wright suggests we need to ask,

> How are these concepts used and contested by differently positioned actors who draw on local, national and global links in unequal relations of power? How is the context framed by implicit practices and rules – or do actors challenge, stretch or interpret them as part of the contest too? In a flow of events, who has the power to define? How do they prevent other ways of thinking about these concepts from being heard? How do they manage to make their meanings stick, and use institutions to make their meanings authoritative? With what material outcomes?
>
> (Wright, 1998: 9)

This politicized view of culture presents cultural values as part and parcel of the political process. Culture is no longer absolute, nor separate nor non-negotiable but part of the enduring competition for power by conflicting interests. Like all forms of politics, the struggle for control of the meaning to be attributed to significant cultural issues is susceptible to mechanisms that, through systems of rules productive of outcomes, enable the routinization of that struggle. Particularly where cultural values stand in the way of scientific and economic advance, a method such as bioethics that brings the cultural opposition from behind the barricades to the negotiating table is likely to be welcomed by governments desirous of solutions. Once at the table that opposition may find that its values have no special ontological status but are converted and defined merely as counters in what may be termed a moral currency that can be traded in the political marketplace. As we have seen in Chapter 6, if the moral currency is dependable enough, if it has international credibility and if it is generally convertible then the exchange of values the currency facilitates contributes to the operation of a global moral economy. The currency must of course remain neutral and must not be seen as having its own interest but as constituting the impartial means for achieving fruitful moral trade. If a currency is successful in encouraging global moral trade, then this is likely to be marked by an increase in its own political value.

Cultural conflict in the EU over the moral status of the human embryo is not new. As early as March 1989 a parliamentary resolution on the ethical and legal problems of genetic engineering had called for legislation prohibiting gene transfer to human germ line cells and had defined the legal status of the human embryo in order to provide une-quivocal protection of genetic identity (European Parliament, 1989). Other resolutions followed in 1993, 1997 and 1998 opposing cloning of the human embryo, supporting the Council of Europe's *Convention on Human Rights and Biomedicine* and calling on Member States to intro-duce a legally binding ban on the cloning of human beings (European Parliament, 1993, 1997, 1998). Meanwhile, in response to these and other cultural concerns about the implications of scientific advance for the status of the human body, the EGE was developing a role as a body with the authority to make ethical pronouncements on such issues. Originally established in 1991 as the Group of Advisers to the European Commission on the Ethical Implications of Biotechnology, the EGE currently comprises 15 members who advise the President of the Commission on how to include the ethical values of European society in the Community's scientific and technological policymaking. This 'long' history of the European Group on Ethics has played a key role in helping to establish it as a trusted and respected institution in the political decision-making process.

Its advice is contained in 'Opinions' on particular topics produced by the EGE, usually in response to a request from the President, but on occasions acting on its own initiative. The EGE has produced Opinions on gene therapy (1994), prenatal genetic diagnosis (1995), the patenting of inventions involving elements of human origin (1996), the genetic modification of animals (1996), cloning (1997), human-tissue banking (1998), patenting inventions involving human stem cells (2002), human stem cell research (2002), genetic testing in the workplace (2003), umbilical cord blood banking (2004) and the ethics review of hESC FP7 research projects (2007), among others (Group of Advisers on the Ethical Implications of Biotechnology, 1994, 1995, 1996a, 1996b, 1997; EGE 1998a, 2002a, 2002b, 2003, 2004, 2007). Its procedure is to draw on international and European legal instruments and the work of national ethics committees, and then to outline its own views. In so doing, the EGE sees itself as establishing fundamental ethical principles in response to the fact that 'in spite of powers having remained mainly at national level in matters of ethics, the free European market is not sufficient to satisfy the requirements of European society. It thus falls to the Community authorities to take account of the public's ethical con-cerns, even if they are at variance with certain economic and financial interests underlying the functioning of the market' (EGE, 2001, p. 10).

In effect, the Group sees itself as the guardian of the rights of civil society by enabling 'the Community authorities, which are responsible for regulating the market, to take better account of the aspirations of the public in the various aspects of their lives: as consumers, workers, parents, patients etc.' (EGE, 2001, p. 12). In this role, probably its most significant contribution to the political culture of the EU has been its drafting, at the request of the President of the European Commission, the European Charter of Fundamental Rights for European Citizens, which was adopted in December 2000 by the EU Summit of Heads of State and Government (EGE, 2000b).

In 1998, the EGE moved beyond its role as an external and occasional (if influential) consultant to the Commission to the more substantial one of political broker when the Commission's Research Directorate faced difficulties in formulating an agenda on health biotechnology for FP5. On the one hand, science and industry wanted more support for Europe's health biotechnology sector. On the other, the bruising political experience over GM foods and crops had shown the European Parliament that investment in biotechnology was an issue to which it should be sensitive. To help resolve this dilemma, the EGE was asked to produce an Opinion on the ethical aspects of human embryo research – on which stem cell research depends (EGE, 1998b).

Undoubtedly influenced by that Opinion, the FP5 1998–2002 research agenda was defined to exclude any 'research activity which modifies or is intended to modify the genetic heritage of human beings by altera-tion of germ cells' and any 'research activity understood in the sense of the term "cloning", with the aim of replacing a germ or embryo cell nucleus with that of the cell of any individual, a cell from an embryo or a cell coming from a later stage of development to the human embryo' (European Council, 1999). At the same time, and again on the recom-mendation of the EGE Opinion that FP5 should introduce the principle of ethical review, Article 7 of the Decision concerning FP5 stated that 'all research activities conducted pursuant to the fifth framework pro-gramme shall be carried out in compliance with fundamental ethical principles' (European Parliament, 2001c, 85). The EGE had become a policy broker.

The success of the EGE in its use of formal ethical debate as a vehi-cle for facilitating policy decisions in areas of cultural conflict over science and the status of the human body was also apparent in the contentious policy domain of patenting. What parts of the body can or cannot be owned and why? Here, the eventual emergence of Directive 98/44/EC on the legal protection of biotechnological inventions was dependent on a number of interventions by the EGE and its insistence that ethical considerations form part of the legislation itself (Salter, 2006a). As a consequence of its 1996 Opinion *The Patenting of Inventions Involving Elements of Human Origin*, the Group of Advisers on the Ethical Implications of Biotechnology not only won a formal place for ethics in the directive but also succeeded in stipulating that the following areas were to be excluded on the basis of public order and morality (Group of Advisers on the Ethical Implications of Biotechnology, 1996b):

(a) processes for cloning human beings;
(b) processes for modifying the germ line genetic identify of human beings;
(c) uses of human embryos for industrial or commercial purposes;
(d) processes for modifying the genetic identity of animals which are likely to cause them suffering without any substantial medical benefit to man or animal, and also animals resulting from this process.

   The role of the EGE (created in November 1997 as the successor to the Group of Advisers) was formally incorporated into the directive as both recital and article. Thus Recital 19 notes that account has been taken of Opinion No. 8 of the Group of Advisers, and Recital 44 and Article 7 state that the EGE 'evaluates all aspects of biotechnology ... including where it is consulted on patent law' (European Parliament and Council of Ministers, 1998).

National pressures and ethical boundaries

The agreement on the biotechnology directive consolidated and legitimized the position of the EGE as a bioethical mechanism for the treatment of difficult cultural issues surrounding the status of the human body. To that extent the EGE became an important player in the evolution of the moral economy of the EU. In Chapter 6 we saw how the moral economy of hESC science can be analysed in terms of units derived from the building blocks of the science itself that are simultaneously scientific and ethical objects (Table 7.1).

   These units provide the basis for cultural trading between political authorities and the search for a workable compromise. In the case of the EU, Member States bring national positions to EU-level negotiations over the moral economy of hESC science: positions that have been defined at the Member State level through ethical debate and then enshrined in national legislation. In 2003, a European Commission

Table 7.1   The components of cultural trading

- embryo source
  - aborted
  - IVF supernumerary
  - non-IVF donated
  - cloned
- embryo creation date
- embryo age
- hESC line origin
- hESC line creation date
- hESC line research purpose

survey of the 15 Member State policies concerning hESCs and FP6 at the time of the debate revealed five major regulatory positions when expressed in terms of the above units, positions from which we can assume their cultural bargaining would begin (Table 7.2):[1]

1. Prohibition of procurement of hESCs from human embryos
2. Prohibition of procurement of hESCs but allowing importation
3. Allowing procurement of hESCs from supernumerary human embryos
4. Prohibition of creation of human embryos for research purposes including cloning
5. Allowing the creation of human embryos for research purposes, including cloning

At one end of the continuum of regulatory positions, the UK's Human Fertilization and Embryology (Research Purposes) Regulation 2001 extended the HFE Act 1990 to permit the use of embryos, regardless of source, in research in order to increase knowledge about serious diseases and their treatment. At the other, the Irish constitution of 1937 (as amended in 1983) provided that 'the State acknowledges the right to life of the unborn and, with due regard to the equal right to life of the mother, guarantees in its laws to respect, and as far as practicable, by its laws to defend and vindicate that right' (European Commission, 2003b, p. 42). As Table 7.2 shows, between the UK and Irish regulations were a variety of positions and non-positions constructed by states seeking to reconcile conflicting cultural pressures from civil society, science and industry. So as we saw in Chapters 4, 5 and 6 in an attempt to remove the human embryo from the political equation (and/or to distance themselves from the act of embryo destruction necessary for hESC creation), Germany, and also Austria and Denmark have allowed the

Table 7.2 Regulations in EU Member States regarding hESC research (March 2003)

Type of regulatory control	Austria	Belgium	Denmark	Germany	Spain	Finland	France	Greece	Ireland	Italy	Luxembourg	Netherlands	Portugal	Sweden	UK
Prohibition of human embryo research									X						
Prohibition of the procurement of ESCs from human embryos					X		X		X						
Prohibition of procurement of ESCs from human embryos but allowing by law for importation of human ESCs	X		X	X											
Allowing for the procurement of human ESCs from supernumerary embryos by law					X			X				X		X	X
Prohibition of the creation of human embryos for research purposes by law or by ratification of the Council of Europe's Convention on Human Rights and Biomedicine	X		X		X	X	X	X	X				X	X	X
Allowing for the creation of human embryos for ESC procurement by law															X
No specific legislation regarding human embryo research				X						X	X		X		

Source: European Commission, 2003b: Table 1 (amended).

importation of hESC lines while internally prohibiting their procurement from human embryos. Commenting on the ethical contortions involved in the policy of hESC lines importation, the EGE noted 'a tendency to accept double morality where there is no coherence between different positions adopted by country'. It continued: 'One could expect that to consider research on human embryos to derive stem cells as unethical, might imply the prohibition of the import for research of embryonic stem cells derived from human embryos as well as of the use of potential therapeutic applications resulting from such research, which is not always the case' (EGE, 2002, para 1.21). In the difficult world of human embryo politics, 'double morality' may well be the ethical price that must be paid for a political compromise.

As the agenda-setting debate for FP6 got under way, and as the move advanced to establish health as the first priority for genomics and biotechnology, the cultural tensions inherent in the range of Member State policy positions became evident. As we have seen, the political atmosphere was already well heated, particularly in the European Parliament. In this context, the publication on 16 August 2000 of the UK Department of Health's *Stem Cell Research: Medical Progress with Responsibility* not only raised the political temperature dramatically but also succeeded in focusing the European Parliament's attention by recommending that research should be permitted on human embryos for therapeutic purposes, including for SCNT: a very permissive approach when compared to other EU countries. National cultural differences now had a specific target issue and, with the FP6 awaiting approval and contingent on parliamentary support, a powerful institutional vehicle for registering conflicting cultural values. (Decisions on FPs have to be made by co-decision between the Council and Parliament.)

The subsequent EU debate about hESC research and FP6 constituted a struggle for control of the political discourse and thus of the policy agenda. Within the debate emerged a range of more detailed positions, again constructed from the units of cultural trading listed in Table 7.1 and using combinations outlined in Table 7.3. Each ethical cell in the matrices can be regarded as potential agenda-setting territory and thus as a political resource that through the operation of the moral economy may be exchanged for, or coupled with, other cells as trading takes place within the political discourse.

Institutional struggle and political discourse

Given the diversity of Member State policy positions on the moral status of the human embryo, it was to be expected that the United Kingdom's report advocating greater freedoms for hESC research would be provocative and ethically challenging. Responding to that report, the European Parliament passed a resolution on 7 September 2000 opposed to both reproductive and therapeutic cloning. Therapeutic cloning (Table 7.3, cells 13–16) was seen as 'irreversibly crossing a boundary in

Table7.3   Combinations of ethical units in the
moral economy of hESC science

ESC Conditional date of creation	Embryo Conditional date of creation	
	No	Yes
	Donated embryo	
No	1	2
Yes	3	4
	Supernumerary embryo	
No	5	6
Yes	7	8
	Aborted embryo	
No	9	10
Yes	11	12
	Cloned embryo	
No	13	14
Yes	15	16

research norms' and as contrary to public policy as adopted by the EU (European Parliament, 2000). Much of the debate was couched in emotive and categorical terms with little suggestion from the opponents of hESC research that negotiation was either possible or proper. In its report *Ethical Aspects of Human Stem Cell Research and Use*, published two months later, the EGE took a more sophisticated view and began the process of establishing an ethical continuum of types of human embryo and hESC research, using the kinds of criteria employed in Table 7.3, and suggesting that some criteria are more acceptable than others (EGE, 2000b). Although it regarded spare (supernumerary) embryos as an appropriate source for stem cell research, in an interesting conditional formulation, it deemed 'the creation of embryos with gametes donated for the purpose of stem cell procurement [as] ethically unacceptable, when spare embryos represent a ready alternative' (EGE, 2000a, para 2.7; Table 7.3, cells 5–8, plus a conditional acceptance of cells 1–4). Meanwhile, 'the creation of embryos by somatic cell nuclear transfer for research on stem cell therapy would be *premature*', since there are alternative sources (EGE, 2000a, para 2.7, emphasis added; Table 7.3, cells 13–16). Embedded in this discourse are notions, first, of embryo status contingent upon source and, second, of ethics as a developmental process that moves from 'premature' to, presumably, mature.

The EGE report signalled an attempt by some actors involved in constructing the political narrative to change the debate from one characterized by static and opposing ethical positions to one in which successive refinements of position were normal and negotiation possible. In other words, it sought to establish the basic requirements of a moral economy. As time drew nearer for Parliament to consider the Commission's FP6 proposal, and as the critics of human embryo research made it clear that they would use this as an opportunity for expressing their opposition, the objective need for negotiating room increased. Although the subsequent parliamentary debate on the First Reading of the proposal in November 2001 suggests that little had changed, and that categorical statements of broad ethical positions were still the norm, the amendments incorporated into the proposal prove otherwise. The amendments meant that FP6 would not fund 'research activity aiming at human cloning for reproductive purposes' or 'the creation of embryos for research purposes including somatic cell nuclear transfer' (therapeutic cloning – Table 7.3, cells 13–16). However, it would fund (and here is the compromise) 'research on "supernumerary" early-stage (i.e., up to 14 days) human embryos (embryos genuinely created for the treatment of infertility so as to increase the success rate of IVF but no longer needed for that purpose and when destined for destruction)' (European Parliament, 2001a, Article 3; Table 7.3, cells 5–8).

The success of these amendments indicates that there is a subtext of covert political negotiation and cultural trading around the ethical units of embryo source and embryo age (up to 14 days). (The latter is, of course, elsewhere described as the 'pre-embryo', an important political category in the long-running UK embryo debate (Mulkay, 1997, 30–2; Spallone, 1999)). Under pressure from the conflicting political constituencies of FP6, the political discourse was beginning to evolve and to suggest that some types of embryos are ethically more important than others. In an attempt to facilitate this evolution and as part of the search for a way through the thickets of the ethical debate, in December 2001 the European Parliament set up the Temporary Committee on Human Genetics and Other New Technologies of Modern Medicine to report on the ethical, legal, economic and social implications of human genetics. In the event, its activities served to stimulate the involvement of new civil society policy networks in the discussion and legitimized the inclusion of fresh ethical dimensions. As the debate on its final report on 29 November 2001 demonstrates, ethical collisions in the parliamentary arena were at this stage more achievable than were compromise positions (European Parliament, 2001b).

In contrast to this, in the separate arena of the EGE a moral economy was clearly at work. The expert agenda of human embryo research was experiencing a further process of ethical refinement in response to scientific and industrial demands for greater regulatory protection of their hESC investments. By 2002, there had been 500 patent applications worldwide referring to ESCs – of which one quarter had been granted – but the EU's position on whether patents on hESCs should or could be granted under the conditions of its 1998 patent directive remained unresolved (EGE, 2002, para 1.16). The directive was clear that industrial and commercial exploitation of human embryos is excluded from patenting but unclear about the patentability of cells obtained from embryos, regardless of embryo source (EGE, 2002, para 1.21). Reflecting on this issue, the EGE stated its opinion that 'patenting of inventions allowing the transformation of unmodified stem cells from human embryonic origin into genetically modified stem cell lines or specific differentiated stem cell lines for specific therapeutic or other uses, is ethically acceptable as long as the inventions fulfil the criteria of patentability' (EGE, 2002, para 2.5). However, this liberalization of the ethics of patenting was balanced by the EGE's view on therapeutic cloning. Drawing on its earlier Opinion *Ethical Aspects of Human Stem Cell Research and Use*, the EGE called for 'a cautious approach, excluding the patentability of the process of creation of a human embryo by cloning for stem cells' (EGE, 2002, para 2.5).

It is clear that at this stage the search for practical ethical solutions to cultural conflict around hESCs and thus the activation of the moral economy was progressing much more swiftly in the expert arena of the bioethicists than in the parliamentary and Council arenas of the politicians. Nonetheless it was in the latter two arenas that a way forward had to be found if FP6 was to be funded. Institutional struggle was about to begin in earnest. In an interesting and, in the view of the opponents of human embryo research, challenging manoeuvre, at the Second Reading of the FP6 proposal in June 2002 Parliament voted through the overarching FP and transferred the issue of the criteria for embryo and hESC research to the process for approval of the relevant Specific Research Programme (European Parliament and the European Council, 2002). This meant that the Parliament was not directly involved in the decision-making because under EU procedures the Specific Programme details are a 'technical issue' and can be decided on by the Council without the agreement of Parliament. However, the advantage gained by this institutional move appeared to be short lived when in September of the same year, under pressure from Austria, Italy, Germany and Ireland,

the Council decided on a package of measures in response to the opposition concerns. This reiterated the ban on therapeutic cloning research
and, furthermore, stipulated that there should be a moratorium on the
EU funding of human embryo and hESC research until December 2003;
a report on hESC research as the basis for an inter-institutional seminar
on bioethics; and, taking into account the seminar's outcome, further
guidelines on the principles that should guide Community funding of
such research to be produced by December 2003 (European Council,
2002). (In an interesting concession to the United Kingdom's pro-hESC
research stance, the moratorium explicitly did not include 'banked or
isolated human human embryonic stem cells in culture'.)

The increasing salience of novel modes of organized ethical engagement (Temporary Committee on Human Genetics, EGE Opinions, inter-
institutional seminar, development of ethical funding guidelines, ethical
review of projects) is an indicator of the intensifying search for practical
means that would enable a moral economy to work and thus promote
the inclusion of cultural factors in the EU's transnational governance
of the life sciences. In its March 2003 progress report *Life Sciences and
Biotechnology – A Strategy for Europe*, the Commission observed:

> Public authorities at large have to take into consideration concerns
> about the conditions under which fundamental choices are made in
> this field [of life sciences]. For its part the Commission is committed
> to ensuring that the ethical, legal, social and wider cultural aspects,
> as well as the different underlying ways of thinking, are taken into
> account at the earliest possible stage in Community-funded research.
> In particular, the issues of *human cloning and human embryonic stem
> cell research* have provoked intense public and political debate. Ethical
> and social debate must continue to be a natural part of the research
> and development process involving society as much as possible.
>
> (European Commission, 2003a, p. 3, emphasis in original)

However, the widespread recognition that cultural values are a legitimate component of the transnational governance of the life sciences
did not readily lead to a parallel acceptance of the new mechanisms
for the resolution of cultural conflict. Following the Commission's
exhaustive report on hESC research and the inter-institutional seminar
drawing on its findings in April 2003 (European Commission, 2003b,
2003c), the terms and constituency of the debate were undoubtedly
enhanced – but so also was the difficulty of finding a sustainable

compromise position. Cultural trading was taking place, but no agreement could yet be reached.

Driven to a search for the lowest common ethical denominator, in June 2003 the Commission proposed a set of ethical guidelines that included the selection of embryos for research using criteria of embryo source and date of embryo creation. Community funding was to be restricted to the derivation of hESC lines 'from human embryos created as a result of medically-assisted in vitro fertilisation designed to induce pregnancy and were no longer to be used for that purpose' (supernumerary embryos) and created before 27 June 2002, the date of approval of the overarching FP6 (Table 7.3, cells 6 and 8) (European Commission, 2003e, pp. 4–5). Unsurprisingly, international scientists objected strongly to the date of embryo creation criterion because of what they saw as its impact on the freedom and quality of their research (Research Europe, 2003). Under the terms of the consultation procedure, the European Parliament debated the Commission's proposal in November 2003 and, agreeing with the scientific view, not only removed the 27 June 2002 restriction but also enlarged the embryo source criterion to include those produced by spontaneous or therapeutic abortion (Table 7.3, cells 9–12) as well as supernumerary embryos from IVF treatment (European Parliament, 2003a). This amendment in turn proved unacceptable to the Council with the result that on 31 December 2003 the moratorium on hESC research expired with no agreement on the principles that should guide Community funding of that research. Under the EU's Comitology rules, if Council fails to make a decision on a Commission proposal, the Commission can implement it. By default, therefore, the criteria contained in the European Parliament and Council Decision of 27 June 2002 and the Council Decision of 30 September 2002 in respect of the Specific Research Programme remained in place. HESC research using therapeutic cloning (Table 7.3, cells 13–16) could not be funded, research based on supernumary embryos (Table 7.3, cells 5–8) could, and the position of research using donated and aborted embryos (Table 7.3, cells 1–4 and 9–12) as the source remained unresolved. FP6 went ahead on this basis and commenced on 1 January 2004. The politics of hESC science had taken the EU right up to the wire.

The volatility of the continuing cultural politics of hESC research and FP6 is manifest both in the constantly shifting mosaic of ethical units in the political discourse and in the absence of any pattern in the institutional struggles between Commission, Council and Parliament. A stable moral economy had yet to emerge. As different configurations of Table 7.3's ethical units came to the fore at different times, so the institutions would change their positions.

A further destabilizing influence was the engagement between the political chemistry, networks and forces at work in the policy domain of FP6 and that of the neighbouring policy field of human tissues where a directive was being considered. The proposal for a directive setting quality and safety standards in relation to human tissues and cells began its progress through the EU's legislative machinery on 19 June 2002 (8 days before the approval of FP6) and immediately became the focus of a conflict not about ethics as such (though this formed part of the debate) but, more important, about what ethics could legitimately be included in the discussion. In this respect, it became a test case for determining what role ethics should have in this policymaking domain and thus whether it was legitimate for a moral economy to exist at all in certain policy fields.

The opponents of human embryo research saw the directive as the means for implementing a pre-emptive strike against the pro-hESC lobby. If the ethical units of Table 7.3 could be inserted into the directive as a block on hESC research, Member States would be obliged to implement it at the national level. The activities of their scientists would thus be curtailed regardless of the outcome of the conflict in the FP6 policy domain. However, the idea that difficult ethical issues should be incorporated into the directive did not resonate well with the culture of the sponsoring Commission Directorate Health and Consumer Protection, which saw the business of setting standards for the donation, procurement, testing, processing, storage and distribution of human tissues and cells as a largely technical exercise with ethics making a facilitative rather than a challenging contribution to the implementation of an existing policy agenda. In 1998, the European Group on Ethics had produced its report *Ethical Aspects of Human Tissue Banking*, which dealt with ethical issues such as the protection of health, the integrity of the human body, informed consent and the protection of identity, and these issues were happily incorporated into the first draft of the proposal for a directive (EGE, 1998a; European Commission, 2002). In an aside, the proposal noted that 'germ cells, foetal cells/tissues and embryonic stem cells pose particular ethical problems', that 'there is no consensus among Member States upon which basic harmonised decisions at EU level can be taken with regard to their use or prohibition' and that 'the proposal does not interfere with decisions made by Member States concerning the use or non-use of any specific type of human cells, including germ cells and embryonic stem cells' (European Commission, 2002, pp. 5–6).

This hands-off approach was abruptly challenged by the Committee on the Environment, Public Health and Consumer Policy in its report to the Parliament as background to the First Reading of the proposal. (It is no coincidence that the rapporteur for the Committee was Peter Liese, a Catholic Christian Democrat Member of the European Parliament (MEP) who was also active in the hESC and FP6 arena.) Here it was proposed that Member States should at least prohibit research on human cloning for reproductive purposes or to supply stem cells, including by means of the transfer of somatic cell nuclei; that no tissues or cells derived from human embryos should be used for transplantation; and that cloned human embryos and human/animal hybrid embryos produced by cloning should be excluded as sources of material for transplant (European Parliament, 2003b, amendments 14, 30, 51). The subsequent acceptance of these amendments by Parliament shifted the focus of the ethical debate from utilitarian values concerned with the details of directive implementation to fundamental values that questioned parts of the science on which the directive was, or might be, based. To counter this, in revising the proposal, the Commission used the interesting tactic of defining some ethics as appropriate to the directive and others not. Therefore, although the directive was able to accept ethical provisions related to the anonymity of donors and non-profit procurement, the Commission argued that other provisions (notably those concerned with human embryos) fell 'outside the scope of Article 152 of the Treaty, which provides for public health protection and *not for the implementation of ethical objectives'* (European Parliament, 2003c: 7, emphasis added). In other words, it argued that only a very restricted kind of moral economy was appropriate to this issue.

In the Second Reading of the directive by Parliament, this selective approach to the role of ethics was sustained, and the opponents of hESC research were obliged to accept a compromise amendment that protected the rights of Member States to ban or restrict the use of ESCs and stipulated that, where used, they should be subject to the directive's provisions for the protection of public health (European Parliament and the European Council, 2004). The human tissues Directive 2003/23/EC was accepted, and the attempt to use it as a vehicle for blocking hESC research was foiled.

Stabilizing cultural conflict: FP7 and the bureaucratization of ethics

The instability of the cultural conflict over FP6 and hESC science left the Commission in an exposed position insofar as particular decisions on the funding of FP6 hESC projects might be challenged by Parliament. To deal with this situation, the Commission introduced a new bureaucratic device, the *Procedural Modalities for Research Activities Involving Banked or Isolate Human Embryonic Stem Cells in Culture to Be Funded under Council Decision 2002/834/EC* (European Commission, 2003d), to form part of the Ethical Review of 'sensitive' FP6 applications. Under this procedure, all hESC projects were obliged to have an ethical review and then to be approved on a case-by-case basis by a special regulatory committee composed of Member States. The ethical review was also to ensure compliance of the project with the legislation of the nation in which its research was to be conducted. Although FP6 rules did not exclude hESC derivation from supernumerary embryos, or other research on such embryos, in practice only proposals using existing hESC lines were submitted and funded – an implicit recognition of what was politically acceptable.

The implementation of the procedural modalities on hESC science under FP6 succeeded in stabilizing the cultural conflict so far as that particular funding programme was concerned within a bureaucratic procedure that was regarded as acceptable by the main players. (Eighteen proposals involving hESCs were reviewed out of a total 855 projects that underwent ethical review in FP6 (Fitzgerald, 2006).) The interesting political question, however, was whether that bureaucratic device would be a sustainable instrument of governance over time. The enlargement of the EU and the addition of ten new Member States on 1 May 2004 changed the pattern of the cultural domain. The national positions of the Member States on hESC science and the human embryo, and therefore the cultural values they were likely to seek to pursue with regard to EU policies in that field, are listed in Table 7.4.

Table 7.4 Regulations in EU Member States regarding hESC research (2005)

Belgium, UK and Sweden allow therapeutic cloning

Denmark, Finland, France, Greece Spain and Netherlands allow the derivation of new hESC lines from supernumerary IVF embryos

Estonia, Hungary, Latvia and Slovenia have no specific regulations on hESC, but allow some research on supernumerary IVF embryos

Germany and Italy have regulations which restrict hESC research: scientists cannot drive new hESC, but can import them. In Germany, these cells have to have been derived before 1 January 2002

Ireland and Slovakia prohibit procurement of hESCs from human embryos

Austria, Lithuania and Poland have legislation prohibiting hESC research

Portugal, Luxembourg, Malta and Cyprus have no specific legislation

The issue of the moral status of the human embryo was not going to disappear. On 10 March 2005, the European Parliament demonstrated its continuing interest in the subject by adopting a resolution calling for a ban on the trade in human egg cells. Embedded within the resolution was the call for funding of ESC research from national budgets and for EU funding to be used for alternatives such as somatic stem cell and umbilical cord stem cell research (European Parliament, 2005). Five months later, parliamentary opposition to the prospect of hESC funding in FP7 emerged in the form of a letter from a group of 73 MEPs to the European Commission President, Manuel Barroso, calling on the Commission to respect the principle of subsidiarity in the matter: Member States should decide if they want to fund research in the human embryo or not (EurActiv.com, 2005).

However, although early signs suggested a repeat of the complex manoeuvrings over FP6, underpinned by a range of opposing value positions, this did not happen. At the First Reading of the proposed FP7 programme on 15 June 2005, Parliament reiterated the de facto position resulting from the lengthy FP6 debates. The following types of research should not be publicly financed: human cloning for reproductive purposes, research intended to modify the genetic heritage of human beings that could make such changes heritable and research intended to create human embryos solely for the purpose of research or for stem cell procurement, including by means of SCNT ('therapeutic cloning'). In addition, MEPs agreed that research on the use of human stem cells, both adult and embryonic, may be financed 'depending on the contents of the scientific proposal and the legal framework of the Member State(s) involved' (European Parliament, 2006). Thereafter, the debate never broadened to include the wide range of ethical units contained in Table 7.3 (type of embryo, date of creation) that was characterized in the FP6 discussions. (The exception to this was an amendment demanding that financing of research should be limited to ESC lines created before 31 December 2003 – rejected by Parliament at the First Reading.) Effectively, the previous pragmatic consensus, reinforced by its incorporation in a bureaucratic form that was judged to be successful, held firm. Problems were experienced at the Council with opposition from Austria, Germany and Italy plus the new entrants Lithuania, Malta, Poland and Slovakia. However, Germany and Italy were won over by the following political/ethical compromise:

The European Commission will continue with the current practice and will not submit to the Regulatory Committee proposals for projects which include research activities intended to destroy human embryos, including for the procurement of stem cells. The exclusion of funding of this step of research will not prevent Community funding of subsequent steps involving human embryonic stem cells.

(Research Fortnight, 2006)

With this compromise, FP7 was agreed.

Conclusions

The therapeutic promise of hESC has generated a global competition for the control of its social, scientific and industrial future that is increasing in intensity. Countries are investing in the basic research necessary to develop the field, re-examining their regulatory arrangements, and seeking to attract transnational life sciences companies. But they do not operate in a cultural vacuum. Elements in their civil societies may draw on a variety of cultural values to support or oppose what is officially regarded as being in the national scientific or industrial interest. To the extent that these cultural pressures are problematic, a cultural politics is generated, characterized by a moral economy in which the trading of values facilitates negotiation and compromise.

In the case of the EU, these pressures are localized through the interaction of Member State positions in the context of the EU's institutions and procedures. Member State cultures as revealed in legislative form are not static but are themselves responsive to the international context. Thus, for example, in July 2004 the French parliament banned reproductive human cloning as a 'crime against the human species' but postponed its ban on the use of supernumerary embryos for embryo research, thus allowing certain types of hESC research to continue (Channelnewsasia, 2004). With less equivocation, in September 2004 the new Spanish socialist government announced that it would permit hESC research and viewed therapeutic cloning as 'an open matter' (Yahoo! News, 2004; *The Scientist*, 2004). As Member States change their positions, the matrix of forces at work in the Commission, Council and Parliament also shifts to create a continuing volatility in the balance between the ethical units of the moral economy.

However, some arenas in that economy are more volatile than others. Whereas the public debates of the European Parliament on hESCs and FP6 were usually characterized by the stark presentation of conflicting cultural positions, in the expert arena of bioethics the search for compromise of ethical equations has generated a quite different political style characterized by reason, flexibility and adaptation. In the former, the cultural politics were raw and challenging, but in the latter the explicit search for political utility has necessitated the development of the rules and procedures that can contribute to a practical outcome. Cultural politics in the EU is therefore operating at two levels to accommodate the otherwise incompatible requirements of (a) the unchanging legitimacy of particular value positions and (b) the need for those positions to be negotiable. As the application of the ethical units of Table 7.3 to the political discourse of both levels has illustrated, a range of finely graded value positions on hESC research existed that constitute the currency for biopolitical trading. Although this trade would be denied at the public level, the evidence of the political discourse is that such trading indeed occurred.

The bureaucratic incorporation of the results of that trading in the 'Procedural modalities' governing the application of hESC projects for FP funding has succeeded in stabilizing the moral economy at a particular juncture (at least for the time being). Despite attempts by the opponents of hESC research to open up the issue during the discussions over FP7, it never really took off. On the other hand, neither have the supporters of hESC research made any further gains. Any research that involves the destruction of the human embryo will not be funded: FP7 will only fund the 'subsequent steps' in the hESC research process – that is, research on stem cell lines created by projects funded by other agencies.

For the future, and as the therapeutic applications of hESC research become more evident, the prospect is one of a continuing engagement between the policy domains of hESC science and human tissues. This will be overlaid with a continuing cultural struggle for control of the EU's emergent new methods for the transnational governance of science. There will then be an objective need for the clarification of the role of ethics in the political discourse and of what Gottweis terms 'the ethics infrastructure' (Gottweis, 2003). National and transnational cultural groupings are becoming increasingly sophisticated in the formation and presentation of ethical arguments in this field and will require a parallel improvement in the way in which ethics is used as a form of political currency and exchange in the moral economy. Attempts to exclude ethical issues from the EU's politics as occurred in the case of the human tissue directive are likely to prove counterproductive because of the established and growing cultural pressures on the policymaking apparatus of the EU.

This analysis suggests that institutionalized modes of ethics engagement will become a political technology that constitutes a permanent feature of the new cultural politics as mechanisms are sought to enable the refining, manipulating, resolving and legitimating of cultural differences through the trading of values in an authoritative language and setting. Such modes are likely to continue to operate in parallel to the formal procedures of Commission, Council and Parliament in an attempt to offset the ponderous limits of these institutions to deal with cultural politics. This chapter has noted the politically functional contribution of the EGE not only to the lubrication of the ethical interaction through its elaboration of fresh ethical distinctions and perspectives but also to the facilitation of decision-making through the judicious use of its claim to impartiality. The steady development of the EGE to the status of a trusted and respected institution was key in giving it substantial weight in the hESC decision-making process. Thus, bioethicists are emerging as a new epistemic power group capable of brokering difficult cultural deals at both the national and international levels, and their inclusion in the transnational governance of the EU is part of a global process (Salter and Jones, 2005). As the EU case has shown, in the hESC field they can enable the interrogation of ethical options, and thus the refinement of the political currency of the moral economy, through the investment of ethical significance in such characteristics as the source, date of creation, age and research purpose of the embryo or hESC. Over time, and if they are functionally successful, we may find that the command of ethical as opposed to scientific expertise elevates bioethicists to the status of what may be termed 'the new technocrats' of transnational scientific governance.

8
Contested Governance: Uncertainty and Standardization in Research and Patenting

In the previous chapters, we have mapped some of the diversities and uncertainties of the hESC field. We have seen the range of cultural reactions to embryo research and the volatility of national political responses to the challenge of integrating a stem cell research programme into society. We have also seen that, to an extent, national regulatory regimes appear to be converging and losing some of their volatile character. We saw in Chapter 5 that the rise of bioethical governance has produced a certain regularization of the international debates, identifying particular moral demarcation points in the new hESC technologies – SCNT but not reproductive cloning, for example, or the acceptability of discarded reproductive embryos but not the manufacture of embryos for research – and ordering national legislative responses accordingly. Many nations have moved, over time, towards the liberal end of the regulatory spectrum. While the previous two chapters focused on the global regularizing effects of bioethical governance, in this chapter we focus on another force for global convergence and governance – the demand for standardization. Standardization processes are essential for any scientific field to develop and are applicable to all stages of the knowledge-production process from the basic science to the market product. They are required for collaboration, for the performance of laboratory tasks, for the assessment of research results, for peer review, for the development of IP claims and for clinical applications. Standardization is necessary in all scientific fields and in the commercialization of scientific results, but in hESC research, the political volatility of the research and the divergence of national regulations introduce an extra dimension. Attempts to develop global technical standardization are entwined with global bioethical standardization. Scientific professional bodies are recognizing that international research collaboration will necessitate internationally agreed-upon bioethical standards if scientists are to comply with the national legislation that governs both their own and their partners' research and gain access to international funding (*Nature* Editorial, 2006).

In this chapter, we focus on two nodal points in the knowledge economy of hESC science where the drive for global technical standards is interwoven with a drive for global bioethical standards: the UKSCB (standardization of the scientific process) and, closely associated with this, the patenting of hESCs (standardization of the market process). How do these two domains of governance interlink and what are the implications of our analysis for the development of global governance?

Standardization, science and global governance

Standardization is a central process of all scientific practice, and one of the major demarcators of scientific from non-scientific knowledge. Standardization is important in science because it creates the conditions for stable comparison and the interoperability of technical elements. Scientific discovery is impossible without agreed-upon measures, protocols, classificatory systems and technical benchmarks shared by laboratories working in the same research field (Timmermans and Berg, 2003). Scientists require shared definitions and classificatory systems if they are to move from the particularity of their laboratory language and culture and work with other scientists in other sites. Standards bind communities of practice together across space. Stable classification systems ensure that concepts and definitions are the same in every geographical location and cultural context, allowing predictable communication and shared understanding and practice (Timmermans et al., 1998).

Standardization of laboratory protocols is also essential for scientific practice. Laboratory objects like stem cell lines are inherently prone to artifactual distortions and contaminations. Different laboratory practices will produce different outcomes and objects and scientific management of this volatility requires more 'black boxing' (Latour, 1987), more standardization of technologies, processes and outputs, to ensure the credibility and stability of a discovery. Agreed-upon laboratory protocols produce predictability, stability and easily reproducible results. Standardization is also essential for collaboration; if a research programme is to be carried out in numerous sites, the technical conditions in each site must be the same to eliminate artifactual 'noise' and make results comparable. (Eriksson and Webster, 2008).

Standards are not merely technical artefacts, however, and the process of agreeing standards is intertwined with broader issues of scientific governance and the negotiation of social relations. In many domains, standards touch on issues of public good (e.g., safety standards) and ethics. As we will discuss in detail, technical standards for stem cell line derivation are a complex example of this ethical aspect of standardization. Moreover, standards emerge from the political process generated by negotiation, debate and compromise between interested parties, including factions within scientific communities, state agencies, community groups and market interests, and the outcome is not socially neutral (Bowker and Star, 1999). As Schepel notes,

> [Standards] bring to the table economic, political, moral and technical arguments and ultimately arrive at a solution that will to some extent hurt some groups and in some degree benefit others.... Standardization is a microcosm of social practices, political preferences, economic calculation, scientific necessity and professional judgement.
>
> (2005, p. 6)

Hence, there are strong incentives for scientific and professional bodies to take the initiative in creating standards. Their configuration can determine to a certain degree who has the upper hand in a research field or a market, and this configuration, once made, tends to persist. As Bowker and Star (1999) note, standards have significant inertia and are expensive and time consuming to change, so the interests encoded in a set of standards may provide benefits for a long time.

Finally, questions of standardization are important for our investigation because technical standards play an increasingly important role in global governance. Standards are usually negotiated between private and public bodies, and states necessarily take a strong interest in technical matters with public good implications (Schepel, 2005). At the same time, the growing importance of global standard setting as a form of governance reflects the difficulties for state regulatory power to manage issues that exceed national boundaries. A major effect of globalization is the decentering of state power and its weakening as a central regulatory coordinator in the face of risk management, industrial developments and scientific innovations that do not observe national boundaries. As one commentator puts it,

> Markets take on a greater prominence at both international and domestic levels of government, but not because of a philosophical

decision to cede power to the private sector....Rather, the market and private actors are more prominent because they can approach problems without the limits of arbitrary, territorial boundaries imposed on them.

(Aman, 2001, p. 391)

Naturally aware of this, states seek to engage proactively with the process of globalization in order to influence emerging forms of governance to their own national advantage. As part of this political arena, standardization processes that involve public bodies with non-state agencies and expertise become vehicles for international regulation through professional affiliation, negotiation and consensus building, rather than through top down, state-driven law. Standardization processes and risk management are forms of social steering that align the regulatory interests of states with the research and market needs of science.

The importance of standards networks in globalised risk regulation is their capacity to internalise and renegotiate the boundaries between science and politics, to tie expert knowledge to local professional judgements, institutional structures, social relationships and economic conditions.

(Schepel, 2005, p. 28)

Frequently they relate intimately to formal institutions of governance such as the EU and the World Intellectual Property Organisation (WIPO) that construct legally binding forms of standardization.

Research standardization and the UKSCB

The hESC field is very poorly standardized compared to more established fields of biomedical research (e.g., genomics). As a very new field, which has attracted large numbers of scientists, established working laboratories and funding streams very recently, many of the basic aspects of classification, culture protocols and laboratory handling are not yet established. Eriksson and Webster (2008) note that stem cells present certain problems for stable classification precisely because, by definition, they lack a biological identity. All that the varieties of stem cells have in common is their potential to differentiate into dedicated cell types, and to date there is no single biomarker that can be used to securely classify all types of hESCs. They also present difficulties of handling, storage, transfer between laboratories and laboratory procedure.

The quality that makes them epistemologically and therapeutically valuable, their pluripotency, is fragile and requires considerable husbandry if it is to be retained across sites.

> The handling of stem cell lines, whether for clinical or research purposes, requires considerable care since, when manipulated and passaged in vitro, they can be prone to subtle changes which may not only damage their ability to replicate as stem cells but also cause a loss of their original capacity to differentiate into different cell types. Whilst it is likely to prove an extremely hard or even an impossible task to standardise the cells themselves, the framework of procedures and conditions under which they are cultured, preserved and characterized can be carefully controlled and documented so that these may be reproduced in the recipient's laboratory.
>
> (Healy et al., 2005, 1982)

This material fragility, combined with the ethical volatility of hESC lines, means that many social and technical interests converge around the issues of standardization. As Eriksson and Webster put it, 'The question asked by regulators as well as scientists is how the socio-technical 'quality' of ESCs can be stabilized and thereby provide assurance that stem cell lines will function the way they are supposed/hoped to do' (2008, p. 59).

The UKSCB has emerged as a key institution in managing these standardization issues at both a national and an international level. As we saw in Chapter 2, the bank is the central institution managing stem cell quality and supply in the United Kingdom. All British researchers are obliged to deposit a sample of any hESC line derived in the United Kingdom in the bank, although in practice some commercial researchers have so far avoided compliance. While researchers may access stem cells from elsewhere, they must nevertheless inform the UKSCB's steering committee. In this sense the bank is constituted as an 'obligatory passage point' (Latour, 1988) for all stem cell research in the United Kingdom, a point through which all researchers must pass to carry out their work. The bank also accepts lines from international depositors and makes them available to international researchers. This is one way in which the bank exceeds a national remit. By acting as an international passage point and clearing house for stem cell researchers worldwide, it necessarily extends standardization practices beyond national boundaries. Researchers worldwide are required to meet the bank's requirements regarding the quality of the cells and also the bioethical procedures and rationales set out by the bank's Steering Committee. This bioethical component is discussed in greater detail later.

The bank has also extended its remit through its role in the initiation and coordination of the International Stem Cell Initiative (ISCI), a transnational network of stem cell scientists designed to promote the standardization of the basic science. The aim of the Initiative is to compare and biologically characterize the majority of ESC lines available worldwide and place the data in the public domain. Participant laboratories in the initiative come from Australia, the United Kingdom, the United States, Canada, Sweden, Finland, Japan, Israel, the Netherlands, and the Czech Republic. The ISCI Steering Committee states that the project is 'critical for progress in the field to understand the similarities and the differences between the various isolates, so that research results from different laboratories can be compared in a meaningful fashion' (Steering Committee of the International Stem Cell Initiative, 2005, p. 795).

The UKSCB is providing technical coordination to the project. Each project uses standardized methodologies, cell cultures and reference reagents provided by the bank, and deposits its resulting lines with the bank, which encodes them and forwards them to expert laboratories for investigation and benchmarking. This gives it a powerful position as a key institution in the emerging collaborative relations between the participating laboratories made possible by exactly the processes of standardization that are in train. As the Steering Committee notes, the ISCI paves the way for truly global collaboration.

> The opportunities and challenges in the field are sufficiently complex, and the resources required to address them so substantial, that the field will need to rely not only on individual investigator project grants but also on large, international collaborations along the lines of the Human Genome project.
>
> (Steering Committee of the International
> Stem Cell Initiative, 2005, p. 797)

In this way the UKSCB has positioned itself as a key broker and negotiator across the transnational networks of the international scientific community. As the global leader in brokering technical standardization, and as the access point to well-characterized and stabilized stem cell stock, it is a creator and enforcer of the material standards that will mediate relationships between laboratories in the field.

The UKSCB is also developing forms of standardization that will mediate between more diverse groups. Human tissue banks are institutions that must manage complex and often competing social values and the often-divergent interests of a variety of social actors. Human tissues like hESC lines are the repositories of often incommensurable values – ontological significance, community significance, market value, therapeutic usefulness and research utility (Waldby and Mitchell, 2006). These disparities in value are nowhere more marked than in tissues arising from human embryos. Hence, as the leading tissue bank in the stem cell area, the UKSCB has a complex bioethical and biopolitical brief – to reconcile successfully the interests of embryo donors, civil society, researchers, industry and clinical users.

The bank's Steering Committee has an explicit brief to maintain both technical and ethical standards for the pubic good. In making such decisions, the Steering Committee necessarily relies on forms of standardization that attempt such reconciliation. As Bowker and Star (1999) note, institutions that deal with complex, irreducible ethical issues, for example, hospitals, whose staff must take frequent life and death decisions on behalf of patients, rely on processes of codification and standardization to avoid constant recourse to internal and external debate and conflict. Codification depoliticizes the issues involved, making them more procedural and less conflict ridden. 'Algorithms for codification do not resolve the moral questions involved, although they may obscure them ... when a seemingly neutral ... mechanism is substituted for ethical conflict about the content of the forms, the moral debate is partly erased' (Bowker and Star, 1999, p. 24).

The primary technology used by the bank to manage these kinds of conflicts is its criteria for donor information and consent. Any researcher wishing to deposit cell lines with the bank must demonstrate that their lines have been derived from embryos donated under a codified set of conditions. The Code of Practice states:

Each gamete provider must consent in writing to the following:

1. to the use of embryos created using their gametes in the research project for the derivation of stem cell lines
2. that they understand that a sample of any stem cell line will be deposited in the UK Stem Cell Bank and that the derived stem cell lines may be used in other research projects
3. that they are under no obligation to take part in the study and that a decision not to participate will not alter the treatment that they would normally receive

4. that they understand that they have a right to withdraw their consent without giving any reason, at any stage until the gametes and/or
embryos have been used for research
5. that they understand that any cell line derived from their donated
gametes/embryos may eventually be used for treatment purposes ...
in the future
6. that they understand that cell lines or discoveries made using them
may be patented and used for commercial purposes, but that the
donor will not benefit financially from this
7. whether they agree to be contacted in the future ... in relation to
confirmed test results performed on stem cell lines that are of direct
relevance to their own, their family's or public health.

(Code of Practice for the use of Human
Stem Cell Lines, 2006, p. 16)

The conditions themselves represent possible points of conflict
between donors and other interested parties. Some of them are designed
to protect donors from undue pressure from a research community
closely aligned with the IVF industry. Hence, point 1 is intended to
ensure that donors fully understand the kind of research their embryos
will be used for. Points 3 and 4 are intended to ensure that potential
donors don't feel that their fertility treatment hinges on donation.
Point 7 concerns the bank's duty of care regarding the donors' genetic
health. Points 2 and 5 make clear the destination of the embryos. They
inform the donors that, once donated, they become research objects
at the disposal of laboratories and the UKSCB, and they may eventually become therapeutic objects used by clinicians and patients. Along
with point 4, they clarify that the donation is a non-reversible process,
and once given, the embryos cannot be returned. In that sense, they
negotiate the relationship between the donors and the scientific community and ensure that the research community has clear rights of disposal. Point 6 is perhaps the most contentious, and mediates relations
between donors and commercial interests in stem cell research. HESC
lines are patentable entities in the UK, and this clause in the consent
form is designed to deliver unencumbered IP rights to the laboratory
recipient of the embryo donation.

Through the dissemination and enforcement of the donor consent
conditions, the bank has standardized complex ethical and social negotiations that might otherwise impede research progress. These relations
are not limited to the British nation. For international depositors, the
same consent requirements apply, creating impetus for international

laboratories to harmonize their consent requirements, and the social values they encode, with the UKSCB. The participants in the ISCI must certify that their lines were derived using similar ethical guidelines (Steering Committee of the International Stem Cell Initiative, 2005). Hence, the UKSCB uses its bioethical guidelines as a gatekeeping device. Laboratories wishing to get access to its considerable resources, and to the collaborative networks developing around it, must demonstrate that the ethical provenance of their stem cell lines conforms to those deemed appropriate by the bank's Steering Committee. Clearly, this creates pressure for harmonization of consent procedures among a number of disparate laboratories and the IVF clinics they use to procure embryos. Over time, it seems likely that its consent requirements will create pressure for an international harmonization of consent procedures at a national regulatory level, particularly among nations whose scientists work closely with UK scientists. The bank already has a formal understanding with the US National Institutes of Health, so that hESC research given ethical clearance by the NIH is accepted as cleared by the bank Steering Committee as well, and it seems likely that such memoranda of understanding will multiply as a way of streamlining the collaborative process.

At the same time, it is important to acknowledge the *limits* of bioethical standardization. While in the controlled space of the laboratory, biological standards can be made relatively stable, bioethical standards are much more volatile in their effects. As we have seen in the preceding chapters, they must operate in a complex social field where local culture, hierarchies and values may undermine the ideal social relations of autonomy, informed consent and decisional freedom both assumed in and created by the consent procedure. Bharadwaj and Glasner's (2004) ethnographic study of embryo donation in India suggests the extent to which consent procedures can be followed to the letter, but the social context in which they operate nullifies many of the aims of the procedure. Hence, they found that Indian clinics would routinely offer free IVF for couples who were prepared to donate embryos for research. They note that in the context of the Indian social stress on fertility, the expense of IVF treatment and the relative lack of value placed on embryos, the offer of free treatment constitutes overwhelming pressure, particularly on the female partner, who must otherwise bear the stigma of childlessness. Glasner notes elsewhere (Glasner, 2005) that this local complexity of the social relations of consent, particularly in cases of embryos procured outside Western Europe, will need to be taken into account by the UKSCB in its assessment of the provenance of stem cell lines, an accounting that would necessarily mean moving beyond procedural approaches to assessment. These are the kinds of social complexities that standardization processes are designed to mute and simplify.

Patenting standardization

Domains of scientific governance are rarely separate from, or impervious to, their governance neighbours. Although the UKSCB is an influential player in the standardization of basic hESC research, it acts within an overall governance context that may facilitate or constrain the global significance of its standardization procedures. In this section we explore how the governance domain of patenting, an arena of standardization apparently quite distinct from that of research, can nonetheless impact on how the standardization of research progresses.

In their 2005 report *Intellectual Property as an Economic Asset: Key Areas in Valuation and Exploitation*, the EPO and the OECD argue that in the global knowledge economy an increasing share of the market value of firms derives from their intellectual assets. They continue,

> As firms shift to more open models of innovation based on col-laboration and external sourcing of knowledge, they are exploiting patents not only by incorporating protected inventions into new products, process and services, but also by licensing them to other firms or public research organisations (PROs). Moreover, they are using patents as bargaining chips in negotiations and as a means of attracting external financing from banks, venture capitalists and other sources.
>
> (EPO and OECD, 2005, p. 3)

Pursuing this economic logic, IPR are regarded as an essential compo-nent of this kind of economy because they commodify the intangible capital of knowledge, generate value and facilitate trading. Without IPR and, in particular, patent protection, emerging markets such as stem cell science would find it difficult to develop because the tangible prod-uct has yet to appear and economic value is embedded in the potential application of the knowledge. This problem is particularly acute in high-tech and research-based Small to Medium Enterprises (SMEs) for whom their IPR is their main asset. The economic significance of patents is further enhanced by the need for new forms of knowledge to compete for attention in an increasingly global venture capital market with its own clear demands. Investors, often institutional investors, make their decisions in the light of the patents held by companies (Florida and Samber, 1999; Haemmig, 2003). For capitalization of a new knowledge market to occur, then, investors need to be reassured that the value of the knowledge, as opposed to the value of the eventual product, is in the hands of the company concerned (Zeller, 2005, p. 17). Investors are likely to be particularly sensitive to the patenting issue in high-risk areas such as the early stage development of health biotechnologies where the science is very new and the potential therapies very distant.

As we saw in Chapter 2, among nation states, the United States has pursued the economic logic of IPR most vigorously, leading the drive to establish health biotechnology as a domain of highly protected IP, exemplified by the 1980 US Supreme Court decision on *Diamond* v *Chakrabarty*, when it was ruled that a living organism (in this case a genetically engineered bacterium) could be patented. In general, the Court commented, patents could be granted for 'anything under the sun that is made by man' and in this respect living organisms are not exceptional (Jasanoff, 2005b, p. 49) – a generous view of intellectual ownership. Other states have been less persuaded that the knowledge property generated by the life sciences should be so broadly interpreted. In 2002, Canada's Supreme Court rejected Harvard University's application for a patent on its oncomouse (a mouse with a cancer promoting gene) on the grounds that higher life forms are distinctive and transcend the patenting definition of 'composition of matter' (Nador and Loucaides, 2003, p. 7).

Although individual states may resist the economic logic of patenting so comprehensively embraced by the United States, the national level is not the only, nor necessarily the most critical, political site where the conflict over the appropriate definitions of ownership of the products of the life sciences takes place. Two international bodies have provided a continuing target for political pressure through their attempts to promote the international standardization of biotechnology patenting rules: the WTO's Agreement on TRIPS and the EPO.

TRIPS, created in 1994, sets mandatory harmonizing standards of IP law for two major technologies: digital technology and biotechnology, and has the power to enforce these standards for all members of the WTO (currently 144 states) (Drahos, with Braithwaite, 2002, pp. 10–17). These standards have been consistently contested by two distinct groups in the field of health biotechnology. First, developing nations with burgeoning health manufacture sectors, such as India, Argentina, Brazil and Turkey, have sought to protect their own industries from the force of US and European patents. At their insistence, TRIPS contains a 10-year delay for the institution of pharmaceutical and agricultural chemical patent protection in developing countries. The second group are life scientists themselves. Many are disenchanted with a form of patenting standardization that in their view has the effect of restricting the free flow of scientific information. It should be remembered that the TRIPS agreement was signed at a time in the mid-1990s when the HGP was strongly promoting the open science model (HUGO, 1995). In this context, the issuing of patents on research tools such as the oncomouse granted to Harvard University and Dupont Corporation, and the breast and ovarian cancer gene to the University of Utah, the NIH and Myriad Genetics created the strong suspicion among scientists that 'disproportionate and overlapping patent grants [are] gluing up the research world' (Cornish et al., 2003, p. 19). In their report *Intellectual Property Rights and Genetics* to the UK Department of Health, the authors expressed the fear that as a result of the US-style of approach to patenting in the knowledge market of new health technologies

> [T]he cost of care will increase; that patients will be deprived of access to new techniques and drugs; that research and testing tools will be withheld; that researchers and carers will not share information; that research will become too complicated to enter upon (perhaps because of the so-called 'anti-commons effect' of there being too many right holders); and equally there could be premature commercialisation in the race to get ahead.
>
> (Cornish et al., 2003, p. 19)

Here we can see that the debate about the relationship between patents and economic advance is now emphasising concerns about the relationship between IPR and research ecologies, rather than simply the efficiency of the patents–innovation linkage. Moreover, the political agenda around patenting standardization is broadening as the pure economic arguments (contested though they may be) are supplemented by cross-cutting discourses about rights (e.g., who should have access to the new health technologies) and ethics (what is the moral basis of patenting?). The future global politics of this field will be greatly influenced by the extent to which these broader, cultural factors are able to penetrate the key decision and policymaking arenas.

Reflecting on the sociocultural significance of patenting, Jasanoff suggests that patents order the process of invention in ways that are 'intrinsically political' because their extension 'to new domains alters basic notions of what is a commodity and who can assert ownership over it'. In biotechnology, patents 'have the effect of removing the thing being patented from the category of nature to the category of artifice – a profound metaphysical shift' (Jasanoff, 2005b, p. 204). As the international debate over patenting has progressed, the reservations regarding the universal applicability of the economic logic of patenting have increased, matched by a growing awareness of the range of alternative perspectives that may have a legitimate place in the formation of patenting standards.

Within the TRIPS Agreement, the recognition that non-economic factors may have a proper role to play in patenting policy finds expression in its Article 27 where the Agreement states:

> Members may exclude from patentability, inventions, the prevention within their territory of the commercial exploitation of which is necessary to protect *ordre public* or morality, including to protect human, animal or plant life or health or to avoid serious prejudice to the environment, provided that such exclusion is not made merely because the exploitation is prohibited by law.
>
> (Nuffield Council on Bioethics, 2002, p. 79)

Other permitted exclusions include 'diagnostic, therapeutic and surgical methods for the treatment of humans or animals and plants and animals other than micro-organisms' (paragraph 3). But for these exclusions to have international political impact there would need to be sustained pressure for their activation.

That pressure has emerged in large part through a challenge to the assumption that knowledge is a private rather than a public good, and that the function of patents is to maximize and privatize personal or corporate IP revenues. Against this the alternative view has developed that knowledge is a public or communal good and that the use of patenting as a market mechanism is less important than its contribution to the achievement of certain human and cultural rights (for example, health, human dignity, cultural identity) (Drahos, 1999; Drahos, with Braithwaite, 2002). For the supporters of TRIPS and the US style of patenting (bearing in mind that US patent law has no morality clause), it has been important to limit the strength of this challenge by trying to keep human rights and its associated ethical arguments contained in a separate policy silo. However, the HIV/AIDS epidemic and the imperative for developing countries to ensure their population's right of access to medicines by resisting the rise in drug prices that accompanied the implementation of TRIPS forced a reappraisal of this separatist approach (Cullet, 2003). Nor is it simply a developing country issue. Citizens in developed countries are likely to be equally energized if they see patenting laws as depriving them of what they consider to be their health care rights. O'Connor predicts that in the United States the promise of stem cell research of cures for cancer, diabetes and Alzheimer's disease means that the pitch of the ownership battle will rise proportionally to the success rate of the research. He continues: 'The public's claim to reasonable access to any crucial life-saving medical breakthroughs that do arise from stem cell research may well force federal, state or local officials to circumvent the existing political opposition to compulsory licenses in the United States' (O'Connor, 2005, p. 666).

HESC patents and ethical contests

In Europe, the transnational governance challenge created by the clash between the demand for exclusive private rights in hESC science and the communal values of local cultures is evident in the continuing political negotiations over two international agreements for the regulation of patenting in this field: the Council of Europe's 1973 European Patent Convention (EPC) and the EU's 1998 Directive on the legal protection of biotechnological inventions (Directive 98/44/EC). All EU Member States have ratified the EPC (as have several non-EU states, notably Switzerland). In addition, in 1999 the EPO stated that it would use the Directive as a supplement to interpretation of the EPC and included it in its Implementing Regulations (Baldock and Kingsbury, 2000). This is significant because Article 6 of the Directive excludes

(a) processes for cloning human beings
(b) processes for modifying the germ line genetic identify of human beings
(c) uses of human embryos for industrial or commercial purposes
(d) processes for modifying the genetic identity of animals which are likely to cause them suffering without any substantial medical benefit to man or animal, and also animals resulting from this process; and these exclusions have to be incorporated into the patenting legislation of Member States.

This agreement not only consolidated questions of ethics as factors in European patenting decisions but also provided specific guidance on the treatment of patent applications involving the human body, gene sequences, cloning, human embryos and genetic modification. It is therefore no coincidence that 5 months later the University of Edinburgh ran into a considerable political storm when Patent No. EP 0695351 entitled 'Isolation, selection and propagation of animal transgenic stem cells' was granted to it by the EPO. The patent was challenged on the basis that its claims extended to a method of SCNT in 'animals' and that this included 'humans'. Unusually, 14 different opponents registered their objection to the patent on the grounds of *ordre public* (Article 53a of the EPC), including the governments of the Netherlands and Italy. Demonstrations by Greenpeace coupled with national and international press coverage rapidly politicized this part of the stem cell field. When the EPO Opposition Division ruled on the University of Edinburgh patent application in July 2002, it asserted that any claims involving hESCs violated the European Patent Convention's Rule 23d(c) that excludes uses of human embryos for industrial and commercial purposes from patentability (EPO, 2002). The knock-on effect was immediate with patent examiners using the decision as a precedent to reject an application concerning James Thomson's technique for driving primate ESCs from the WARF (Vogel, 2004b). Other applications from the California Institute of Technology on a method to isolate neural stem cells from embryonic tissue and from the University of Bonn on a method to differentiate neural cells from mammalian ES cells also remain unresolved. As a result of appeals from the University of Edinburgh, the decision on the principle of hESC patenting now rests with the EPO's final authority, the Enlarged Board of Appeal, a process which may take several years. As a consequence, the future value of hESC science is impossible to calculate and thus the potential market in which biotechs and venture capitalists might invest remains unknown.

It can be argued, therefore, that the standardization of technical and bioethical elements by the UKSCB and the ISCI in the governance domain of hESC research has been matched by the failure of international standardization in the governance domain of hESC patenting. Governance development in different parts of stem cell knowledge production has progressed at different rates with different responses to the cultural context. In the case of patenting governance, the effect of this situation is marked. The USPTO has, to date, granted 41 patents that claim hESCs in their title and front pages. These include patents on culture methods, differentiated cells derived from hESCs and even hESCs per se. By contrast, the EPO has not granted a single patent that makes direct hESC claims (Porter et al., 2006).

The implications of contested stem cell governance for the work of the UKSCB are considerable. The bank's advancement of stem cell science through the facilitation of standardized knowledge exchange will be restricted to the extent that the governance of the patenting domain inhibits the translation from basic science to therapeutic product. The bank is clear that its sphere of governance does not include IP. When stem cell lines are deposited with the bank ownership remains with the depositor but the bank stipulates that

> A pre-requisite for depositing in the UK Stem Cell Bank is that the owner of the stem cell line signs a Materials Deposition Agreement (MDA) with the Bank agreeing to make the stem cell line available to requestors for research purposes, on terms of access to be negotiated between the depositor and any future requestor in the Materials User Licence (MUL).
>
> (UK Stem Cell Bank, 2007)

The depositor, meanwhile, acts within the UK's Patents Act 1977, as amended to implement Directive 98/44/EC, on the legal protection of biotechnological inventions. The latter stipulates that uses of human embryos for industrial or commercial purposes are not patentable inventions, so, as we saw in Chapter 2, the UK's Intellectual Property Office (IPO) will not grant patents for processes of obtaining stem cells from human embryos. Nor will it grant patents on totipotent cells with the potential to develop into an entire human body because the human body at the various stages of its formation and development is excluded from patentability by Paragraph 39a of Schedule A2 to the Patents Act 1977. However, human embryonic pluripotent stem cells, with which the UKSCB is at present primarily concerned, are patentable because they do not have the potential to develop into an entire human body (UK Intellectual Property Office, 2003). Depositors in the UKSCB can therefore benefit financially from their ownership of the stem cell lines through the licensing of the use of the lines to interested researchers and companies.

Because the process of hESC line creation is not patentable under UK law, UK researchers in this field have a fair degree of freedom in their basic research. There is a compromise in this governance domain between the market values of ownership, the cultural worth of the human embryo and the scientific values regarding free flows of information. This is not so in the United States, the home of permissive patenting law. As we discussed in Chapter 2, the patents granted to the WARF and

Geron on the process of hESC differentiation 'embody one of the strongest possible property claims in the field of stem cells, establishing control at the very root of all possible lineages of cellular differentiation' (Bergman and Graff, 2007). However, even in the absence of an *ordre publique* clause, the WARF patent has generated considerable internal opposition from scientists claiming that their research is being slowed down as a result of the 'anti-commons' effect of the 'patent thicket' generated by WARF and Geron (*Nature* News, 2005; Stem Cell Business News, 2007).

Conclusions

The global knowledge economy of hESC science is emerging within a changing governance context that both helps and hinders its development. Bioeconomy and governance are evolving interdependently, but not necessarily efficiently. Furthermore, the various domains that govern the knowledge production process are developing their modes of standardization at different speeds, influenced by national and international cultural values and political interests.

In this chapter we have examined standardization in the hESC governance domains of basic research and patenting. In the former, the transnational networks of science are collaborating under the aegis of the UKSCB and the ISCI to develop forms of standardization that incorporate and satisfy technical and ethical criteria. The institutional leadership of the UK in this domain is well accepted and the process of governance development regarded as a neutral international activity pursued for the good of science. In patenting, however, the situation is much more complex. Here hESC science is situated within an existing structure of global, regional and national governance characterized by a plethora of technical and cultural conflicts manifest in continuing legal debates about standardization. In this domain, science is one player among many and its governance interests frequently collide with those of the market.

As different governance domains in a common process of knowledge production, research and patenting are politically interlinked: developments in one will have implications for the other. For example, a key part of the strategies of companies such as WARF and Geron is to acquire the benefits of their patenting rights in the global as well as the US marketplace of stem cell science: profit maximization would naturally ensure that this was so. As we have seen, their ability to do so will vary according to the culture of the state or region they are seeking to access and the configuration of political interests therein. If they are success-ful, there will be governance implications for the standardization of human embryonic research led by bodies such as the UKSCB. Research standardization is necessary not only to enable scientific advance but also as a condition of commercialization. If the basic research is subse-quently seen to be flawed then the clinical and therapeutic products that flow from that research will be seen as suspect. At the same time, if the IP regime either discourages scientists through the dense patent-ing of the knowledge production process or fails to reward potential investors through a very cautious definition of what may be patented, then commercialization will be impeded that way. It is therefore sen-sible to be cautious when assessing the significance of standardization in any particular domain of hESC governance.

Conclusions: Towards the Global Politics of Stem Cell Research

When it was announced on 20 November 2007 that the journals *Cell* and *Science* were going to publish two papers on iPS, they were immediately hailed as historic breakthroughs in human stem cell research. They appeared to benefit from the progress in hESC research without the need to deal with the multifold bioethical issues involved in hESC research (Gawrylewski, 2007). The groups of Shinya Yamanaka at the University of Kyoto, who is also a senior investigator at the Gladstone Institute of Cardiovascular Disease (GICD) at the University of California, San Francisco, and James Thomson in Wisconsin had used genes to programme human cells so that they had all the characteristics of hESCs – but without being derived from human embryos (Takahashi et al., 2007; Yu et al., 2007).

With iPS cells on the horizon, religious groups and institutions in many countries immediately announced the end of hESC research as a scientific option. Several of the scientists previously involved in hESC research also expressed relief that after all the difficulties and political battles about hESC and cloning research, a new alternative in human stem cell research with vast potentials now seemed to be taking shape (Cyranoski, 2007).

The discussion about the implications of iPS research for the future of human stem cell research had just started when the WARF began to file patents applications in the field of iPS technology. In Japan, the Education, Science and Technology Ministry declared that it would subsidize research into iPS cells, and the Japan Science and Technology Agency (JST), an independent administrative agency, announced that it had dispatched full-time IP experts to Kyoto University to allow for the smooth acquisition of patents based on research findings. As the *Yomiuri*

Shimbun, a Japanese daily, reported about the government's view of iPS research:

> The ministry hopes Japan will become a world leader in regenerative science using iPS cells through its support of participating researchers by such means as providing them with free iPS cells and the implementation of the national network. ... The ministry believes it is essential that Japan unites regenerative medicine experts from across the nation in the study of that field, if the nation wants to keep its nose ahead of U.S. and European research teams that have deep pockets.
>
> (Yomiuri, 2007a, 2007b)

In many respects, iPS technology seemed to be an elegant and ethically non-controversial strategy to address some key targets of regenerative medicine, such as the treatment of neurodegenerative diseases (Gottweis and Minger, 2008). However, closer inspection shows that in the process of creating iPS cells, Yamanaka's team had used a retrovirus to deliver four genes into skin cells taken from a mouse and an adult human. Retroviruses are a type of virus that can also cause mutations in the adult cells, making them cancerous, a clear risk constellation if used in a clinical context. They came to broad public attention with gene therapy trials that had failed and resulted in human casualties. Today retroviruses are widely seen as being too risky to be used in gene therapy clinical trials, the main reason for the worldwide trend to avoid their use (Edelstein et al., 2007, p. 839). In addition to the health risks potentially implied in the therapeutic application of iPS cells, other ethically sensitive scenarios began to be contemplated, such as reprogramming cells to not only pluripotency but also totipotency, or the ability to use iPS cells to create human chimera (Lanza, 2007).

Finally, despite all the hype, the announcement of the death of hESC research seemed to be premature, at least according to the pioneers of iPS research (Hyun et al., 2007). As Alan I. Leshner, the chief executive of the American Association for the Advancement of Science and executive publisher of the journal *Science*, and James A. Thomson argued in an op-ed article in the *Washington Post*:

> Far from vindicating the current U.S. policy of withholding federal funds from many of those working to develop potentially lifesaving embryonic stem cells, recent papers in the journals *Science* and *Cell* described a breakthrough achieved despite political restrictions. In fact, work by both the U.S. and Japanese teams that reprogrammed skin cells depended entirely on previous embryonic stem cell research. ...

Reprogrammed skin cells, incorporating four specific genes known to play a role in making cells versatile, or pluripotent, did seem to behave like embryonic stem cells in mice. But mouse studies frequently fail to pan out in humans, so we don't yet know whether this approach is viable for treating human diseases. We simply cannot invest all our hopes in a single approach. Federal funding is essential for both adult and embryonic stem cell research, even as promising alternatives are beginning to emerge.

(Leshner and Thomson, 2007)

These early moments in the history of iPS technology development are highly reminiscent of what we have discussed in our reconstruction of hESC research on the global scale, and what we think applies on a more general level to the emerging field of regenerative medicine. The recent development of iPS cell research, its scientific modes of operation, the attempt to secure exclusive commercial rights and the highly optimistic representation of iPS cells in the public realm seem to typify observable tendencies in some important areas of regenerative medicine. What in 2006 had begun to materialize as a new basic science approach towards stem cell research in laboratories in the United States and Japan, in 2007 had already turned into a fierce scientific race, ending with the simultaneous publication of key research results on the same day, followed immediately by patent claims. In keeping with the global character of stem cell research, Yamanaka was not only a researcher at the University of Kyoto but had also been recruited in 2006 by the Gladstone Institutes of the University of California, San Francisco, which in 2007 applied for funding from the California Institute of Regenerative Medicine, an offspring of Proposition 71 that had opened up broad public support for stem cell research in California. Although key aspects of iPS research such as the use of retroviruses in gene delivery remained unresolved, iPS technology was nonetheless hailed as a new holy grail: 'This is truly the Holy Grail, to be able to take a few cells from a patient – say a cheek swab or a few skin cells, and turn them into stem cells in the laboratory,' said Robert Lanza, chief scientific officer with Advanced Cell Technology in Worcester, Massachusetts. 'It's a bit like learning how to turn lead into gold.' (2007).

Regenerative medicine and global governance

We have argued that today regenerative medicine with its focus on tissue regeneration has become a global theatre in which the biopolitical forces, technological vectors, corporate strategies and patterns of innovation clearly have transcended the confinements of the nation state and operate in the transnational arena. The dynamics of the development of regenerative medicine seem to be driven by a novel global, socioeconomic, political and scientific-technological constellation. As shown in Chapters 1 and 2, and as also evident in the case of iPS cells, stem cell research and regenerative medicine more generally have turned rapidly from a basic science 'breakthrough' into a target for nations and corporations competing worldwide for influence, economic growth and health policy strategies, and for biological materials such as oocytes or hESC lines. This newly emerging knowledge system has given rise to a new knowledge economy, and its vectors operate simultaneously on a national, a regional and a global level where researchers, policymakers, biological materials, stocks and interests circulate. Vast expectations, hopes and scenarios of transforming health care and whole economies seem to coexist with the basic nature of most of the current research in those fields of stem cell research that internationally receive not only most of the attention but also financial research support, such as in hESC and, most recently, iPS research.

But these new biological strategies and technologies to 'program and reprogram life' have also opened up a new global space of uncertainty for governance. During the 1970s and 1980s, political discourse on genetic engineering and its applications focused on the risks associated with genetic engineering and the various strategies to control risk. Today political-institutional strategies towards stem cell research and cloning are characterized by the extensive consideration of the ethical ambivalences of these lines of research and the exploration of their very meaning, applicability, impact and implications. We discussed in Chapters 3 and 4 how hESC research has turned into a political battlefield in many countries, and how it was constructed, staged and interpreted in different cultures. In the emerging policy scenographies, topics of ethical uncertainty seemed to dominate issues of risk during these debates, probably because any clinical application seemed to be so far away. In this constellation, bioethical reflection and bioethics institutions based less on practices of calculative rationality and more on deliberation and reflection gained importance, as we have shown in Chapters 5 and 6. At the same time, technologies of risk continued to be relevant and have recently strongly surfaced in iPS cell research. Thus, in the newly emerging politics of regenerative medicine, risk *and* uncertainty operate as joint modalities of government. Both bioethics committees and administrative agencies concerned, for example, with the surveillance of patients who have received engineered tissues and with the safety of retroviruses in gene therapy are linked in novel ways in the emerging regime of biomedical regulation.

Not surprisingly, the rise of stem cell and cloning technologies world-wide has been accompanied by a vast mobilization of ethical expertise and an almost hectic movement of ethical institution building. This newly emerging ethics infrastructure has no mandate to make political decisions in the legal-regulatory field, but it sets in motion a process of deliberation, reflection and confession that connects people, science, culture, belief systems, religion and other metanarratives with the daily management of stem cell and cloning technologies. As we saw in Chapters 3 and 4, in the United States as well as in Germany and the United Kingdom, the question of whether hESC research should be conducted at all was a central issue in the political debate. The resulting decisions in these debates are less important than the very procedure of deliberation set in motion. Institutions were created with much care and fanfare, which made possible ordered ways to problematize and define potentially controversial topics of medical research. The various ethics institutions and the mobilization of ethical arguments were not simply focused on 'Yes–No' questions. Complex issues gained attention, such as how to frame 'parental consent' in the context of ES research, under which circumstances 'embryo donations' should take place and what the appropriate setting for 'parental decisions' would be. In the United States, Germany and the United Kingdom, complicated legal constructions were created that ensured no work with hESCs would become possible without making the 'parents' of the 'research embryos' an integral part of the decision-making structure.

The staging of bioethical debate about stem cells and cloning seems to be focused on a set of interrelated questions ranging from definitions of the emergence of human life and individuality, personality and the perils of the slippery slopes of contemporary medical technology. For example, objection to cloning is based not only on theological and/or philosophical definitions of human individuality but also on the more general consideration of desirable social development. But, at a different level, this controversy is also about individualizing strategies in body politics. During the 1960s and 1970s, the abortion debates were polarized between those who advocated the sanctity of life even at earlier stages of embryonic developments, and those who supported the rights of woman to dispose of their embryos and foetuses. In a way, the debate was about something 'negative': should women have the right to terminate their pregnancy – or not? In the first decade of the twenty-first century the debate has taken a positive turn: Do individuals have the right to use their embryonic cells to initiate SCNT and autologous therapy? Do they have the right to use embryo cells from other bodies to initiate treatments of illness? Should somatic cells from humans be reprogrammed to create ES-like cells? While institutional politics has become deeply involved in the deliberation of such issues, the debate has also clearly moved beyond the realm of legislatures and ethics institutions. Patient advocate groups such as those dealing with Parkinson's and Alzheimer's have turned into major political forces. Patients suffering from spinal injury or neurological diseases – some of them prominent – such as the British House of Commons member Anne Bagg or actor Christopher Reeve – became vocal advocates for establishing a liberal regime for stem cell research and cloning.

As we have demonstrated, arguments in support of stem cell research and cloning are habitually advanced in the name of major social values, such as health, medical progress, or modernity. Risk regulation is expected to guarantee a certain level of safety for the new biomedical technologies. At the same time, in parallel with the tendency for patients to take more responsibility for their own health and bodies, individuals and groups have started to claim their rights and position themselves accordingly for political confrontation. In most countries arguments in favour of stem cell research have come under heavy criticism by a number of actors and groups ranging from churches to feminist activists. These groups do not face a monolithic state as their counterpart but instead face ethical positions, arguments and invitations to join bioethics boards. They do not clash with police troops in front of abortion clinics or at demonstrations but instead meet well-known theologians, patient representatives and philosophers in citizen's conferences. In general, bringing in and, to some extent, 'shaping' publics seems to have been a key element in successful hESC governance since the late 1990s.

For the new knowledge economy of stem cell science to emerge, not only regulations and bioethical standards but also the establishment of standards in scientific practice and IP rights were important. As we have seen in Chapter 8 and our discussion of the UKSCB, standardization played a key role in scientific practice as it creates the conditions for stable comparison, interoperability of technical objects, and thus opportunities for collaboration on a national and transnational level. Whereas standardization propels the unity of a scientific field such as stem cell research and enhances exchange and cooperation, patenting seems to be doing the opposite and has caused fragmentation in stem cell science, in particular, by creating division in patenting practice between two key regions of stem cell research, Europe and the United States. At the same time, stiff licensing fees for commercial users of hESCs and also (much lower) fees for non-profit researchers could stifle research, a process that might be repeated in the case of iPS technology if the patent applications filed in 2007 are granted.

Stem cell research and the politics of reprogramming life

Against the background of reorganization in the life sciences, biomedicalization, the identification of ageing populations as a central policy problem of advanced industrial societies and neoliberal health discourse, individuals and groups have begun to contextualize their support for stem cell research and cloning, and, in a more general way, their endorsement of the idea of 'self-regeneration' as a new form of identity politics. Just as 'knowing one's genes' and adapting one's eating habits is an expression of a prudent lifestyle, the routine cloning of autologous cell material or the storage of umbilical cord blood might become integral elements of the responsible management of one's health and body in the near future (Waldby, 2006). Multiple resistances against this trend will continue, but increasingly they also seem to operate as functional elements of the emerging bioethical-dialogical apparatus. Thus, the new ethics discourse and its various forms of institutionalization appear to constitute an effort to link new strategies in the self-government of individuals with needs of postmodern governance under conditions of radical uncertainty.

In summary, then, the difficult negotiation between regenerative medicine and society is characterized by the emergence of a new discourse in life science, novel narrativizations of the relevance and impact of stem cell research for international competitiveness, the dramatic expansion of bioethical discourse as the mediator between science and society, and also by the success and failure of different approaches in governing stem cell science. This politics of stem cell research operates in a global arena. As our book clearly demonstrates, a coherent, proactive, trust-generating approach towards the regulation of hESC research was a key factor mediating between society and science in hESC research and development on the national as well as on the supra/transnational level, such as the EU. Ethically controversial technologies and strategies in hESC research seemed to be better accepted socially as a new and important knowledge system in policy scenographies where regulators and policymakers chose to pursue a proactive style of dealing with new challenges that not only demonstrated awareness of the potential social, ethical and cultural dimensions of stem cell research but also instilled trust in the various engaged publics.

Advances in human stem cell science will continue on the global level and will further probe social, cultural and political sensibilities about 'reprogramming' human cells, with its ethical ambivalences and implied risks. While at the moment iPS cell technology seems to offer a solution for ethically non-controversial stem cell research, once eggs and sperm are generated from pluripotent stem cells and thus the creation of embryos comes into view, iPS research might come full circle and raise precisely the same question that has been central in hESC and IVF research (*Nature* 2008, 913). The global politics of regenerative medicine will mix private and public components and include stakeholders such as companies and patient groups, along with governments and international and transnational organizations. As the case of hESC research demonstrated, despite strong concerns and reservations worldwide, research has continued in most countries and made stem cell research into a highly innovative new field in life sciences. While ethical concerns about hESC research have not disappeared, and it is likely that hESC research will continue into the foreseeable future, novel approaches that combine stem cell research with other key strategies in regenerative medicine – such as gene therapy, the creation of tissue and stem cell banks, and the move of stem cell research into clinical trails and application – will pose new challenges for bioethics and risk governance. The fragmentation of the system of regenerative medicine through patenting creates another deep challenge for the future. Thus, the global politics of stem cell research will continue to be a daunting task for those involved in shaping its agenda and structure.

Notes

Introduction: Stem Cell Research and Global Biopolitics

1. Williams and colleagues found that in 2000 alone, the UK print and electronic media carried 63 stories about embryonic stem cell technologies (Williams, Kitzinger and Henderson, 2003).
2. Immortalization is a technique for growing living tissue in a laboratory. Immortalized cells divide and multiply in vitro without forming organized tissues like organs or veins. Cells converted into an immortalized cell line will, in theory, divide and multiply forever. Immortalized human cell lines have been important technologies in medical research since the 1950s, and today thousands of human tissue cell lines are in use throughout the world (Lock, 2001).

1 Globalization, Stem Cell Markets and National Interests

1. In a much more wide-ranging argument, Cooper (2006) argues that the whole project of post-Fordist economies is the reanimation and revaluation of both organic and inorganic matter, by putting them into speculative play through the technical reconfigurations and market deregulation of the 1970s and 1980s.

> The biotechnological solution to economic limits seems to encapsulate the speculative euphoria of revalorisation at the most intimate of material levels. At the same time that it writes off the inorganic matter consumed and left over by industrial production, Post-Fordism attempts to reanimate the whole of matter – the garbage heap of industrial waste, from cadavers to fossil fuels – within the process of its own self-valorization. (Cooper, 2006, p. 8)

2. The term *motive force* derives from Schumpeter and 'refers to the adoption of a major innovation in a dynamic sector with potential repercussions throughout the economy' (Jessop, 2002, p. 282, fn. 16).
3. The exception here is Italy, which at the time of writing has no state financial or regulatory support for hESC research.
4. www.stemcellforum.org.uk. The member organizations from funding countries are Academy of Finland, Australian National Health and Medical Research Council, Canadian Institute for Health Research, Deutsche Forschungsgemeinschaft (Germany), INSERM (France), Israel Academy of Sciences and Humanities, Juvenile Diabetes Research Foundation International, Netherlands Organisation for Health Research and Development, RIKEN (Japan), Singapore Biomedical Research Council, Swedish Research Council, UK Medical Research Council, US National Institutes of Health and the Swiss National Science Foundation.

5. Initiatives announced by the governor of Wisconsin in November 2004 include

   • the removal of bureaucratic hurdles for faculty members who want to become entrepreneurs;
   • venture capital provided for start-up businesses through the Department of Commerce; a new $134 million HealthStar Interdisciplinary Research Complex near the University of Wisconsin Hospital and clinics dedicated to innovation and rapid transfer of medical science discoveries to clinical applications;
   • a new $132 million research facility at the Medical College of Wisconsin and Children's Hospital to focus on infectious disease control, cardio-vascular illnesses and bioengineering;
   • annual support of $1.5 million for a new Alzheimer's research initiative;
   • investments of $105 million over the next 5 years in research, education and public health efforts at the University of Wisconsin Medical School and the Medical College of Wisconsin to make progress in areas such as regenerative medicine, stem cell research, molecular medicine, neuroscience and cancer research. www.wisgov.state.wi.us, date accessed 31 August 2005.

2 Embryos, Oocytes, Cell Lines: HESC Science and the Human Tissue Market

1. A term used to describe the subculture of cells in culture.
2. For example, the International Stem Cell Initiative involves 15 research groups in nine different countries and uses standardized culture protocols, reagents and reference antibodies to characterize all existing ES lines, 'as if they were all being studied in a World-Wide laboratory' (UK stem cell bank website www.ukstemcellbank.org.uk accessed 16 February 2006).
3. Totipotent cells give rise to both the embryo itself and the placenta and supporting tissue.
4. See http://stemcells.nih.gov/research/registry/.
5. See http://www.isscr.org/science/sclines.htm.
6. Set out in SCUKSCB (2004) Annex 11:'Terms and Conditions for deposition and access of human stem cell lines'.
7. SCUKSCB (2004) Annex 2a *'Donor Information and Consent Form'*.
8. See http://www.ukstemcellbank.org.uk/catalogue.html.
9. Gesetz zum Schutz von Embryonen (*Embryonenschutzgesetz* – ESchG) as promulgated on 13 December 1990, Federal Law Gazette I, p. 2746.
10. Figure derived from the first, second and third reports of the Central Ethics Committee for Stem Cell Research, covering the period July 2002–November 2005.
11. Sexton (2005), extrapolating from Hwang's figures, claims that almost half the young women in Britain would need to donate oocytes simply to treat those with diabetic conditions.
12. In response to the severe shortages of oocytes within Britain, the Human Fertilization and Embryology Authority, the statutory body which regulates fertility medicine and embryonic research in Britain, has reviewed its payment provisions for sperm and oocyte donation and proposed much more generous

financial and in-kind arrangements than it previously permitted. The *SEED Report* argues that donors should be reimbursed all out-of-pocket expenses incurred within the UK in connection with gamete or embryo donation and also be compensated for loss of earnings up to £250 for each 'course' of sperm donation or each cycle of egg donation (HFEA, 2005). They may also receive discounted treatment services in return for supplying gametes for the treatment of others.

13. Personal communication, Donna Dickenson.

3 Global Regulation and Local Policy Narratives: Making Sense of Dolly

1. This section on Italy follows closely Herbert Gottweis, Ingrid Metzler and Erich Griessler, *Defining Human Life – Human Embryonic Stem Cell Resarch between Politics and Ethics*, Paganini EU FP 6 Report 2, Vienna 2007, with Ingrid Metzler responsible for the section on Italy.
2. We are grateful to Byoungsoo Kim for the reconstruction of the Korean case.

4 From Dolly to Therapies? Stem Cell Regulations in the Making I – The United Kingdom and the United States

1. http://www.prolife.org.uk/; http://www.lifeuk.org/, http://www.spuc.org.uk/index.htm.

5 From Dolly to Therapies? Stem Cell Regulations in the Making II – Germany, Italy, Japan and South Korea

1. This section on Italy follows closely Herbert Gottweis, Ingrid Metzler and Erich Griessler, *Defining Human Life – Human Embryonic Stem Cell Resarch between Politics and Ethics*, Paganini EU FP 6 Report 2, Vienna 2007, with Ingrid Metzler responsible for the section on Italy.
2. We are grateful to Byoungsoo Kim for the reconstruction of the Korean case.

7 HESC Science and the Cultural Politics of the EU's Framework Programmes

1. A global profile of these regulatory positions can be found in Chapter 5.

References

Aldhous, P. (2001) 'Can They Rebuild Us?', *Nature*, 410, 622–5.

Aman, A. (2001) 'The Limits of Globalisation and the Future of Administrative Law: From Government to Governance', *Indiana Journal of Global Legal Studies*, 379.

Andrews, L. and Nelkin, D. (2001) *Body Bazaar: The Market for Human Tissue in the Biotechnology Age* (New York: Crown Publishers).

Annas, G. J., Caplan, A. and Elias, S. (1999) 'Stem Cell Politics, Ethics and Medical Progress', *Nature Medicine*, 5(12), 1339–41.

Appadurai, A. (1996) *Modernity at Large: Cultural Dimensions of Globalization* (Minneapolis: University of Minnesota Press).

Asad, T. (ed.) (1973) *Anthropology and the colonial encounter* (London: Ithaca Press).

Asad, T. (ed.) (1979) Anthropology and the analysis of ideology. *Man.* 14: 607–27.

Australian Government (2005) *Legislation Review: Prohibition of Human Cloning Act 2002 and the Research Involving Human Embryos Act 2002* (Canberra, December).

Australian Stem Cell Centre (2005) Australian Stem Cell Centre, www.nscc.edu.au/nscc_home.html (home page), date accessed 31 August 2005.

Baldock, C. and Kingsbury, O. (2000) 'Where Did It Come from and Where Is It Going? The Biotechnology Directive and Its Relationship to the EPC', *Biotechnology Law Report*, 19(1), 7–17.

Barnett, A. and Smith, H. (2006) 'Cruel Cost of the Human Egg Trade', *The Observer*, 30 April, www.observer.guardian.co.uk, date accessed 26 May 2006.

Bauman, Z. (1998) *Globalization: The Human Consequences* (Oxford: Polity).

Beauchamp, T. I. and Childress, J. F. (1989) *Principles of Bioemedical Ethics*, 3rd edn (New York: Oxford University Press).

Bergman, K. and Graff, G. D. (2007) 'The Global Stem Cell Patent Landscape: Implications for Efficient Technology Transfer and Commercial Development', *Nature Biotechnology*, 25, 419–24.

Betta, M. (1995) *Embryonenforschung und Familie. Zur Politik der Reproduktion in Großbritainnien, Italien und der Bundesrepublik* (Frankfurt am Main: Peter Lang).

Bharadwaj, A. and Glasner, P. (2004) 'Spare Embryos and Biotech Futures: Embryonic Stem Cell Research in India', Paper presented at 4S/EASST Conference, Ecole de Mines, 24–9 August, Paris.

Biobusiness (1996), June, 12–13.

BioCentury (1999) 'The Bernstein Report on BioBusiness', 28 June 12.

Biomed Singapore (2000) 'Major Venture Capital Injection of $17 Million into Regenerative Medicine', 11 August, http://www.biomed-singapore.com/bms/sg/en_uk/index/newsroom/pressrelease/2000/major_venture_capital.html, date accessed 27 July 2005.

Biophoenix (2006) *Opportunities in Stem Cell Research and Commercialization: Technology Advances, Regulatory Impact and Key Players* (Coventry, UK: Biophonoenix).

Bishop, J. P. and Jotterand, F. (2006) 'Bioethics as Biopolitics', *Journal of Medicine and Philosophy*, 31, 205–12.

Bonaccorso, M. (2004) 'Making Connections: Family and Relatedness in Clinics of Assisted Conception in Italy', *Modern Italy* 9(1), 59–68.

Bosch, X. (2005) 'Concerns over New EU Ethics Panel', *The Scientist*, 6(1) (4 November), http://www.the-scientist.com/article/display/22822/.

Bosk, C. L. (1999) 'Professional Ethicist Available: Logical, Secular, Friendly', *Daedelus*, 128(4), 47–68.

Bowker, G. and Star, S. (1999) *Sorting Things Out: Classification and Its Consequences* (Cambridge, MA: MIT Press).

Boyle, J. (1996) *Shamans, Software and Spleens: Law and the Construction of the Information Society* (Cambridge, MA: Harvard University Press).

Boyle, J. (2003) 'The Second Enclosure Movement and the Construction of the Public Domain', *Law and Contemporary Problems*, 66(1, 2), 33–74.

Brocke, P. (2002) 'Lebenshilfe-Vorsitzender Robert Antretter warnt eindringlich vor Import embryonaler Stammzellen', *Bundesvereinigung Lebenshilfe, Pressemitteilung*, 29 January.

Brown, N. and Webster, A. (2004) *New Medical Technologies and Society: Reordering Life* (Malden, MA: Polity Press).

Bush, G. W. (2001) Remarks of the President on Stem Cell Research, August 9, Bush Ranch, Crawford, Texas, http://www.whitehouse.gov/news/releases/2001/08/20010809-2.html, date of access 22 September 2001.

Butler, D. (1998) 'Breakthrough Stirs U.S. Embryo Debate', *Nature* 396 (November 12), 104.

Butler, D. and Wadman, M. (1997) 'Calls for Cloning Ban Sell Science Short', *Nature*, 386, 8–9.

California State Senate (2001) SB 253, An Act to Add Article 5 (commencing with Section 125115) to Chapter 1 of Part 5 of Division 106 of the Health and Safety Code, Relating to Medical Research, 15 February.

Callahan, D. (2006) 'Bioethics and Ideology', *Hastings Center Report*, January–February, 3.

Canadian Stem Cell Network (2005) Stem Cell Network, www.stemcellnetwork.ca (home page), date accessed 26 August 2005.

Catenhusen, W. M. (2000) 'Contributions in discussion at Bundesministerium für Bildung und Forschung', in Bundesministerium für Bildung und Forschung (ed.), *Statusseminar: Die Verwendung humaner embryonaler Stammzellen in der Medizin–Perspektiven und Grenzen* (Bonn, Germany).

Cerny, P. (1997) 'Paradoxes of the Competition State: The Dynamics of Political Globalization', *Government and Opposition*, 32(2), 251–74.

Channelnewsasia (2004) 'French Parliament Bans Human Cloning', http://www.channelnewsasia.com/stories/afp_world/view/94729/1/.html.

Chief Medical Officer's Expert Group (CMOEG) (2000) *Stem Cell Research: Medical Progress with Responsibility: Report from the Chief Medical Officer's Expert Group Reviewing the Potential of Developments in Stem Cell Research and Cell Nuclear Replacement to Benefit Human Health* (UK: Department of Health).

China Ministry of Science of Technology and Ministry of Health (2004) 'Guidelines for Research on Human Embryonic Stem Cells', January, http://www.chinaphs.org/bioethics/regulations_&_laws.htm#_toc113106142.

Cibelli, J. B., Kiessling, A. A., Cunniff, K., Richards, C., Lanza, R. P., and West, M. D. (2001) 'Rapid Communication: Somatic Cell Nuclear Transfer in Humans: Pronuclear and Early Embryonic Development', *Journal of Regenerative Medicine*,

2 (November 26), 25–31, http://www.bedfordresearch.org/articles/cibelli_jregenmed.pdf.

Cimons, M. (2001) 'Stem Cell Study Decision Due by Summer', *Los Angeles Times*, March 1.

Cirant, E. (2005) *Non si gioca con la vita. Una posizione laica sulla procreazione assistita* (Roma: Editori Riuniti).

Clarke, A. J., Mamo, L., Fosket, J., Fishman, J. and Shim, J. (2003) 'Biomedicalization: Theorizing Technoscientific Transformations of Health, Illness, and U.S. biomedicine', *American Sociological Review*, 68(2), 161–94.

Clouser, K. D. (1993) 'Bioethics and Philosophy', *Hastings Center Report*, 23(6), S10–S11.

Clouser, K. D. and Gert, B. (1990) 'A Critique of Principlism', *Journal of Medicine and Philosophy*, 15, 219–36.

Code of Practice for the Use of Human Stem Cell Lines (2006) version 3, August.

Cohen, P. (1998) 'Hold the Champagne', *New Scientist*, 14 November, p. 8.

Coleman, W. and Perl, A. (1999) 'Internationalized Policy Environments and Policy Network Analysis', *Political Studies*, 47(4), 691–709.

Comitato Nazionale per la Bioetica (2000) Parere del comitato nazionale per la bioetica sull'impiego terapeutico delle cellule staminali, 27 ottobre 2000. Documenti del Comitato Nazionale per la Bioetica. P. d. C. d. M. D. p. l. i. e. l'editoria (Roma, Istituto Poligrafico e Zecca dello Stato S.p.A).

Cooke, P. (2002) *Knowledge Economies: Clusters, Learning and Cooperative Advantage* (London: Routledge).

Cookson, C. (2005) 'Country Report: China', *Financial Times*, 17 June, 1.

Cooper, House of Commons Debate (2000) 15 December, c 877.

Cooper, M. (2006) 'Resuscitations – Stem Cells and the "Crisis of Old Age"', *Body & Society* 12(1), 1–23.

Cornish, W. R., Llewelyn, M., and Adcock, M. (2003) *Intellectual Property Rights and Genetics: A Study into the Impact and Management of Intellectual Property Rights Within the Health Care Sector* (Cambridge: Public Health Genetics Unit).

Council for International Organizations of Medical Sciences (CIOMS) (2002) *International Ethical Guidelines for Biomedical Research Involving Human Subjects*, http://www.cioms.ch/frame_guidelines_nov_2002.htm.

Crawford, R. (1984) 'A Cultural Account of "health": Control, Release, and the Social Body, in J. McKinlay (ed.), *Issues in the Political Economy of Health Care* (New York/London: Tavistock), 60–106.

Cullet, P. (2003) 'Patents and Medicines: The Relationship between TRIPS and the Human Right to Health', *International Affairs*, 79(1), 139–60.

Cyranoski, D. (2007) 'Race to Mimic Human Embryonic Stem Cells', *Nature*, 450, 462–3.

Dean, M. (1999) *Governmentality: Power and Rule in Modern Society* (London/Thousand Oaks: Sage).

Delavigne, A. and Rozenberg, S. (2002) 'Epidemiology and Prevention of Ovarian Hyperstimulation Syndrome (OHSS): A Review', *Human Reproduction Update*, 8, 559–77.

Department of Health (Great Britain) (2000a) *Government Response to the Recommendations Made in the Chief Medical Officer´s Expert Group Report 'Stem Cell Research: Medical Progress with Responsibility'*, Presented to Parliament by the

Secretary of State for Health by Command of Her Majesty (London: August Cm 4833).

Department of Health (Great Britain) (2000b) *Stem Cell Research: Medical Progress with Responsibility: A Report from the Chief Medical Officer's Expert Group Reviewing the Potential of Developments in Stem Cell Research and Cell Nuclear Replacement to Benefit Human Health,* June (London: Department of Health).

Deutsche Alzheimergesellschaft (2002) *Leitsätze der Deutschen Alzheimergesellschaft zu ethischen Fragesellungen,* verabschiedete auf der Delegiertenversammlung, 16 November, Erkner.

Deutscher Ärztetag (2001) 'Beschlussprotokoll des 104. Deutschen Ärztetags, Ludwigshafen', 22–5 May.

Deutsche Bioschofskonferenz (2001) *Stellungnahme des Sekretärs der Deutschen Bischofskonferenz, Pater Dr. Hans Langendörfer SJ, zu den Empfehlungen der Deutschen Forschungsgemeinschaft zur Forschung mit menschlichen Stammzellen,* 5 May.

Deutsche Bioschofskonferenz (2002) *Der Vorsitzende der Deutschen Bischofskonferenz, Kardinal Karl Lehmann, und der Vorsitzende des Rates der Evangelischen Kirche in Deutschland (EKD), Präses Manfred Kock, zur Entscheidung des Deutschland Bundestages über den Import menschlicher embryonaler Stammzellen,* 30 January.

Deutscher Bundestag (2000) 14 Wahlperiode, March 22.

Deutscher Bundestag (2001) 14 Wahlperiode, Zweiter Zwischenbericht der Enquete-Kommission Recht und Ethik der modernen Medizin, Teilbericht Stammzellenforschung', 12 November.

Deutscher Bundestag (2002a) 14 Wahlperiode, January.

Deutscher Bundestag (2002b) Schlussbericht der Enquete-Kommission, *Recht und Ethik der Modernen Medizin,* 14 Wahlperiode, Drucksache 14/9020, 14.05, Berlin.

Deutsche Forschungsgemeinschaft (2001) *Neue Empfehlungen der DFG zur Forschung mit menschlichen Stammzellen,* 3 May.

Dickenson, D. (2006) 'The Lady Vanishes: What's Missing from the Stem Cell Debate', *Journal of Bioethical Inquiry,* 3, 1–2, 43–54.

Die Welt (2001) 30 January.

Die Zeit (2001) 5 July.

Drahos, P. (1999) 'Intellectual Property and Human Rights', *Intellectual Property Quarterly,* 3, 349–71.

Drahos, P. with Braithwaite, J. (2002) *Information Feudalism: Who Owns the Knowledge Economy?* (London: Earthscan).

DuBose, E. R., Hamel, R. P., and O'Connell, L. J. (1994) *A Matter of Principles: Ferment in US Bioethics* (Valley Forge, PA: Trinity Press International).

Echlin, H. (2005) 'How Much Would You Pay for This?', *The Guardian,* 25 July, 10–11.

Edelstein, M., Mohammad, E., Abedi, R., and Wixon, J. (2007) 'Gene Therapy Clinical Trials Worldwide to 2007 – An Update', *Journal of Gene Medicine,* 9, 833–43.

Edwards, R. G. (2001) 'IVF and the History of Stem Cells', *Nature,* 413, 349–51.

Edwards, S. (2007) 'Geron Anticipates Embryonic Stem Cell Trials in 2008', *Wired Science,* May 11 blog.wired.com/wiredscience/2007/05/geron_anticipat.html, date accessed 12 November 2007.

Eiseman, E. (2000) 'Quick Response: Use of Human Fetal Tissue in Federally Funded Research', in National Bioethics Advisory Commission (ed.), *Ethical Issues in Human Stem Cell Research* (vol. 2) (Rockville, MD), C1–C7.

Elias, P. (2006) 'Singapore Woos Top Scientists', *ABC News*, 29 June, http://abcnews.go.com/Technology/wireStory?id=1835594, date accessed 19 December 2006.

Ericson, R., Barry, D., and Doyle, A. (2000) 'The Moral Hazards of Neoliberalism: Lessons from the Private Insurance Industry', *Economy and* Society, 29(4), 532–58.

Eriksson, L. and Webster, A. (2008) 'Standardising the Unknown: Practicable Pluripotency as Doable Futures', *Science as Culture*, 17(1), 57–69.

Ethics Committee of the American Society for Reproductive Medicine (2000) 'Financial Incentives in Recruitment of Oöcyte Donors', *Fertility and Sterility*, 74(2), 216.

Etzkowitz, H. (2003) 'Innovation in Innovation: The Triple Helix of University-Industry-Government Relations', *Social Science Information*, 42(3), 293–337.

Etzkowitz, H. and Webster, A. (1995) 'Science as Intellectual Property', in S. Jasanoff, G. Markle, J. Petersen, and T. Pinch (eds) *Handbook of Science & Technology Studies* (London: Sage), 480–505.

EurActiv.com (2005) 'MEPs Challenge Stem Cell Research', 22 September, http://www.euractiv.com/en/science/meps-challenge-eu-stem-cell-research/article-144641.

Europa (2007) 'The Framework Programme for Research and Development and Human Embryonic Stem Cell Research', Press release, 29 March (Brussels: European Commission).

European Commission (2002) *Proposal for a Directive of the European Parliament and of the Council on Setting Standards of Quality and Safety for the Donation, Procurement, Testing, Processing, Storage and Distribution of Human Tissues and Cells.* COM(2002)319 final, 19/6/02.

European Commission (2003a) *Life Sciences and Biotechnology – A Strategy for Europe. Progress Report and Future Orientations.* COM(2003)96 final.

European Commission (2003b) *Report on Human Embryonic Stem Cell Research.* SEC(2003)441, 3 April.

European Commission (2003c) 'EU Funding and the Bio-ethics of Stem-Cell Research. Inter-institutional Seminar', News Report, 25 April, http//:www2.europarl.eu.int/omk/sipade2?PUBREF=-//EP//TE.../EN&LEVEL=2&NAV=.

European Commission (2003d) *Procedural Modalities for Research Activities Involving Banked or Isolated Human Embryonic Stem Cells in Culture to Be Funded under Council Decision 2002/834/EC* (Brussels: European Commission), http://ec.europa.eu/research/science-society/pdf/procedural_modalities_en.pdf.

European Commission (2003e). Proposal for a Council Decision amending decision2002/834/EC on the specific programme for research, technological development and demonstration: 'Integrating and strengthening the European research area' (2002–6). COM(2003)390 final.

European Commission (2005) *Social Values, Science and Technology.* Special Eurobarometer (Cited 8 September 2005), http://europa.eu.int/comm/public_opinion/archives/ebs/ebs_225_report_en.pdf.

European Commission (2006) *Report on the Regulation of Reproductive Cell Donation in the European Union - Results of Survey,* http://ec.europa.eu/health/ph_threats/human_substance/documents/tissues_frep_en.pdf.

European Council (1999) 'Council Decision 1999/167/EC adopting a Specific Programme for Research, Technological Development and Demonstration on Quality of Life and Management of Living Resources, 1998–2002', *Official Journal*, L 64(12/3/99), 1–19.

European Council (2002) 'Council Decision 2002/834/EC Adopting a Specific Programme for Research, Technological Development and Demonstration: "Integrating and Strengthening the European Research Area"' (2002–6), *Official Journal* L 294(29/10/02), 1–43.

European Group on Ethics in Science and New Technologies (1998a) *Ethical Aspects of Human Tissue Banking*, Opinion No. 11 to the European Commission.

European Group on Ethics in Science and New Technologies (1998b) *Ethical Aspects of Research Involving the Use of the Human Embryo in the Context of the Fifth Framework Programme*, Opinion No. 12 to the European Commission.

European Group on Ethics in Science and New Technologies (2000a) *Ethical Aspects of Human Stem Cell Research and Use*, Opinion No. 15 to the European Commission.

European Group on Ethics in Science and New Technologies (2000b) *Citizen Rights and New Technologies: A European Challenge* (Brussels: EGE).

European Group on Ethics in Science and New Technologies (2001) *General Report of the Activities of the European Group on Ethics, 1998–2000* (Brussels: European Commission), http://europa.eu.int/comm/european_group_ethics/publications_en.htm.

European Group on Ethics in Science and New Technologies (2002a) *Ethical Aspects of Patenting Inventions Involving Human Stem Cells*, Opinion No. 16 to the European Commission (Brussels: EGE).

European Group on Ethics in Science and New Technologies (2002b) *Ethical Aspects of Human Stem Cell Research and Use*, Opinion No. 15 to the European Commission (Paris: EGE).

European Group on Ethics in Science and New Technologies (2003) *Ethical Aspects of Genetic Testing in the Workplace*, Opinion No. 18 to the European Commission (Paris: EGE).

European Group on Ethics in Science and New Technologies (2004) *Ethical Aspects of Umbilical Cord Banking*, Opinion No. 19 to the European Commission (Paris: EGE).

European Group on Ethics in Science and New Technologies (2007) *Ethics Review of hESC FP7 Projects*, Opinion No. 22 (Paris: EGE).

European Parliament (1989) 'Resolution on the Ethical and Legal Problems of Genetic Engineering', *Official Journal*, C 096(17/4/89), 171.

European Parliament (1993) 'Resolution on the Cloning of the Human Embryo', *Official Journal*, C 315(22/11/93), 224.

European Parliament (1997) 'Resolution on the Cloning of Human Beings', *Official Journal*, C 115(14/4/9), 92.

European Parliament (1998) 'Resolution on the Cloning of Human Beings', *Official Journal*, C 034(2/2/98), 164.

European Parliament (2000) Resolution on Human Cloning PE T 5-0375/2000.

European Parliament (2001a) Proposal for a European Parliament and Council Decision Concerning the Multiannual Framework Programme 2002–6 of the European Community for Research, Technological Development and Demonstration Activities Aimed at Contributing Towards the Creation of the European Research Area. COM(2001)94, 14 November.

European Parliament (2001b) Debates of the European Parliament. Sitting of 29 November 2001. Human Genetics.

European Parliament (2001c) Temporary Committee on Human Genetics and Other New Technologies in Modern Medicine. *Report on the Ethical, Legal, Economic and Social Implications of Human Genetics*, Final A 5-0391/2001, European Parliament Session Document, Brussels.

European Parliament (2003a) European Parliament Resolution on the Proposal for a Council Decision Amending Decision 2002/834/EC on the Specific Programme for Research, Technological Development and Demonstration: 'Integrating and Strengthening the European Research Area (2002–6)'. COM(2003)390, 19 November.

European Parliament, Committee on the Environment, Public Health and Consumer Policy (2003b). *Report on the Proposal for a European Parliament and Council Directive on Setting Standards of Quality and Safety for the Donation, Procurement, Testing, Processing, Storage, and Distribution of Human Tissues and Cells*. A 5-0103/2003. Final.

European Parliament (2003c) *The Legislative Observatory. Procedure: Medicine: Standards of Quality and Safety of Human Tissues and Cells*.

European Parliament (2005) Resolution on the Trade in Human Egg Cells, 10 March, Strasbourg.

European Parliament (2006) European Parliament Resolution on the Proposal for a Decision of the European Parliament and of the Council Concerning the Seventh Framework Programme of the European Community for Research, Technological Development and Demonstration Activities (2007–13) (COM(2005)0119 – C 6-0099/2005-2005/0043(COD)).

European Parliament and Council of Ministers (1998) Directive 98/44/EC on the Legal Protection of Biotechnological Inventions, 6 July, Brussels.

European Parliament and the European Council (2002) 'Decision 1513/2002/EC of the European Parliament and Council Concerning the Sixth Framework Programme of the European Community for Research, Technological Development and Demonstration Activities, Contributing to the Creation of the European Research Area and to Innovation (2002–6)', *Official Journal*, L232(29/8/02), 1–33.

European Parliament and the European Council (2004) 'Directive 2004/23/EC of the European Parliament and of the Council on Setting Standards of Quality and Safety for the Donation, Procurement, Testing, Processing, Preservation, Storage and Distribution of Human Tissues and Cells', *Official Journal*, L102 (31/03/04), 48–58.

European Patent Office (EPO) (2002) '"Edinburgh" Patent Limited after European Patent Office Opposition Hearing', EPO press release, http://www.european-patent-office.org/news/pressrel/2002_07_24_e.htm z, date accessed 16 June 2006. European Patent Office (EPO) and OECD (2005) *Intellectual Property as an Economic Asset: Key Issues in Valuation and Exploitation* (Paris: OECD).

Evangelische Kirche Deutschlands (2001) *Der Schutz menschlicher Embyonen darf nicht eingeschränkt werden. Erklärung des Rates der EKD zur aktuellen bioethischen Debatte*, May 22.

Evans, J. H. (2000) 'A Sociological Account of the Growth of Principlism', *Hastings Center Report*, 30(5), 31–8.

Evans, J. H. (2002) *Playing God: Human Genetic Engineering and the Rationalisation of Public Bioethical Debate* (Chicago: University of Chicago Press).

Feder Ostrov, B. (2002) *The Mercury News*, 23 September.

Feldman, E. A. (2002) *The Ritual of Rights in Japan: Law, Society, and Health Policy* (Cambridge: Cambridge University Press).

Fink, S. (2005) 'Live and Let ... ? Measuring and Explaining the Strictness of Embryo Research Laws in 21 Countries', Third European Consortium for Political Research Conference, Panel 3–7: The Regulation of the Life Sciences, 8–10 September, Budapest.

Fitzgerald, M. (2006) 'Integrating Ethics in EU Research', Presentation. European Commission Research DG, Brussels, http://www.enterprise-ireland.com/NR/rdonlyres/441E83F 5-5FEB-4A 38-8F 19-7907F04EFF92/09EthicalReviewsMaryFitzgerald.pdf.

Flamigni, C. and Mori, M. (2005) *La legge sulla procreazione medicalmente assistita. Paradigmi e confronti* (Milano: il Saggiatore).

Florida, R. and Samber, M. (1999) 'Capital and Creative Destruction: Venture Capital, Technological Change, and Economic Development', in M. Gertler and T. Barnes (eds), *The New Industrial Geography: Regions, Regulations and Institutions* (London: Routledge), 265–91.

Forster, H. and Ramsey, E. (2001). 'The Law Meets Reproductive Technology: The Prospect for Human Cloning', in P. Lauritzen (ed), *Cloning and the Future of Human Embryo Research* (Oxford: Oxford University Press), 201–21.

Foucault, M. (1983) *Der Wille zum Wissen, Sexualität und Wahrheit 1.* Frankfurt a. M.: Suhrkamp (Original work published in 1976).

Foucault, M. (1994) *Überwachen und Strafen. Die Geburt des Gefängnisses.* Frankfurt a. M.: Suhrkamp. (Original work published in 1975).

Fox, R. (1999) 'Is Medical Education Asking Too Much of Bioethics', *Daedelus*, 128, 1–25.

France, L. (2006) 'Passport, Tickets, Suncream, Sperm...' *The Observer*, January 17, www.observer.guardian.co.uk, date accessed 3 February 2006.

Frankel, M. S. (2000) 'In Search of Stem Cell Policy', *Science*, 287, 1397.

Franklin, S. (2007) *Dolly Mixtures: The Remaking of Genealogy* (Durham: Duke University Press).

Gallese, P. and Toldo, M. (1998) 'Il cao Dolly: storia di una notizia', in R. Satolli and F. Terragni (eds), *La clonazione e il suo doppio* (Milano: Garzanti), 19–41.

Galloux, J., Mortensen, A., De Cheveigne, S., Allansdottir, A., Chatjouli, A., and Sakellaris, G. (2002) 'The Institutions of Bioethics', in M. W. Bauer and G. Gaskell (eds), *Biotechnology: The Making of a Global Controversy* (London: Science Museum), 129–48.

Gawrylewski, A. (2007) 'Two Teams Reprogram Skin Cells for Pluripotency', Scientist.com, 20 November, http://www.the-scientist.com/blog/display/53873/, date accessed 22 December 2007.

Gearhart, J. D. (1998) 'New Potential for Human Embryonic Stem Cells', *Science*, 282, 1061–2.

'Gesetz zur Sicherstellung des Embyonenschutzes im Zusammenhang mit der Einfuhr und der Verwendung menschlicher embryonaler Stammzellen', *Stammzellengesetz StGZ, Bundesgesetzblatt Jahrgang* 2002, 1(42), Bonn, 29 June.

Glasner, P. (2005) 'Banking on Immortality? Exploring the Stem Cell Supply Chain from Embryo to Therapeutic Application', *Current Sociology*, 53(2), 355–66.

Gold, E. R. (1996) *Body Parts: Property Rights and the Ownership of Human Biological Materials* (Washington, DC: Georgetown University Press).

Gottweis, H. (1998) *Governing Molecules: The Discursive Politics of Genetic Engineering in Europe and the United States* (Cambridge, MA: MIT Press).

Gottweis, H. (2003) 'Human Embryonic Stem Cells, Cloning, and the Transformation of Biopolitics', in Nico Stehr (ed.) *Biotechnology between Commerce and Civil Society* (New Brunswick, New Jersey: Transaction Books), 239–65.

Gottweis, H. and Triendl, R. (2006), 'South Korean Policy Failure and the Hwang Debacle', *Nature Biotechnology*, 24, 141–3.

Gottweis, H. and Minger, S. (2008) 'iPS Cells and the Politics of Promise', *Nature Biotechnology*, 26, 271–2.

Gottweis, H. and Petersen, A. (2008) *Biobanks: Governance in Comparative Perspective* (London: Routledge).

Gough, K. (1968) 'New Proposals for Anthropologists', *Current Anthropology*, 9, 403–7.

Government, UK (1996–97) Government [British] Response to the Fifth Report of the House of Commons Select Committee on Science and Technology (1996–7) Session (Cm 3815).

Gross, M. (2003) 'Stem Cells Divide the World', *BIOforum Europe*, 5, 2–3, www.gitverlag.com, date accessed 19 May 2006.

Group of Advisers on the Ethical Implications of Biotechnology (1994) *Gene Therapy*, Opinion No. 4 (Brussels: Group of Advisers).

Group of Advisers on the Ethical Implications of Biotechnology (1995) *Ethical Aspects of Prenatal Diagnosis*, Opinion No. 5 (Brussels: Group of Advisers).

Group of Advisers on the Ethical Implications of Biotechnology (1996a) *Ethical Aspects of the Genetic Modification of Animals*, Opinion No. 7 (Brussels: Group of Advisers).

Group of Advisers on the Ethical Implications of Biotechnology (1996b) *The Patenting of Inventions Involving Elements of Human Origin*, Opinion No. 8 (Brussels: Group of Advisers).

Group of Advisers on the Ethical Implications of Biotechnology (1997) *The Cloning Techniques*, Opinion No. 9 (Brussels: Group of Advisers).

Guardian (2001a) 22 January.

Guardian (2001b) 27 February.

Guardian (2002) 1 March.

Haas, P. M. (2001) 'Policy Knowledge: Epistemic Communities', *International Encyclopedia of the Social and Behavioral Sciences*, 17, 11578–86.

Haemmig, M. (2003) *The Globalisation of Venture Capital: A Management Study of International Venture Capital Firms* (Bern, Stuttgart: Verlag Paul Haupt).

Hall, S. S. (2002) 'President's Bioethics Council Delivers', *Science*, 297(July 19), 322–4.

Hamilton, J. (2002) 'What Are the Costs?', *Stanford Magazine*, 7 January, http://www.stanfordalumni.org/news/magazine/, date accessed 24 January 2006.

Han, J. K. (2006) 'Responsibilities of the Government and Political Circle', Paper Presented at Korean Biotechnology Watch Coalition Forum, Seoul, Korea, 18 January.

Hay, C. (2004) 'Re-stating Politics, Re-politicising the State: Neo-liberalism, Economic Imperatives and the Rise of the Competition State', *Political Quarterly*, 75, 1, 38–50.

Healy, L., Hunt, C., Young, L., and Stacey, G. (2005) 'The UK Stem Cell Bank: Its Role As a Public Research Resource Centre Providing Access to Well-Characterised Seed Stocks of Human Stem Cell Lines', *Advanced Drug Delivery Reviews*, 57(13), 1981–8.

Hedgecoe, A. (2004) 'Critical Bioethics: Beyond the Social Science Critique of Applied Ethics', *Bioethics*, 18(2), 120–43.

Herper, M. (2001) 'Hold Off on Investing in Stem Cells', *Forbes*, 13 August, http://www.forbes.com/2001/08/13/0813steminvest.html, date accessed 4 November 2004.

HGAC (1998a) *Cloning Issues in Reproduction, Science, and Medicine*, January, London.

HGAC (1998b) *Cloning Issues in Reproduction, Science, and Medicine*, December, London.

Hoffman, J. P. and Johnson, S. M. (2005) 'Attitudes toward Abortion among Religious Traditions in the United States: Change or Continuity?', *Sociology of Religion*, http://www.findarticles.com/p/articles/mi_m0sor/is_2_66/ai_n14817990.

House of Lords (2002) *Report from the Select Committee on Stem Cell Research*, 27 February 2002 (HL 83(i)).

Human Fertilisation and Embryology Act (1990) (c. 37) (London: Her Majesty's Printing Office).

Human Genome Organisation (HUGO) (1995) *Statement on the Patenting of DNA Sequences*, http://www.who.int/genomics/elsi/regulatory_data/region/international/162/en/index.html, date accessed 10 March 2008.

Human Genome Organisation (HUGO) (1996) *Statement on the Principled Conduct of Genetics Research*, http://www.gene.ucl.ac.uk/hugo/conduct.htm.

Human Genome Organisation (HUGO) (2004) *Statement on Stem Cells*, http://www.eubios.info/hugostem.pdf, date accessed 4 February 2008.

Hunter, J. D. (1994) *Before the Shooting Begins: Searching for Democracy in America's Culture War* (New York: the Free Press).

Hwang, W. S. et al. (2004a) 'Evidence of a Pluripotent Human Embryonic Stem Cell Line Derived from a Cloned Blastocyst', *Science Online*, February 13, www.sciencemag.org, date accessed 3 April 2006.

Hwang, W. S. et al. (2004b) 'Evidence of a Pluripotent Human Embronic Stem Cell Line Derived from a Cloned Blastocyst', *Science* 303(March 12), 1669–74.

Hwang, W. S. et al. (2005) 'Patient-Specific Embryonic Stem Cells Derived from Human SCNT Blastocysts', *Science* 308 (June 17), 1777–83.

Hwa-Young, T. (2006) 'Ova Donors Demand Compensation from Government', *AsiaNews.it*. www.asianews.it, date accessed 2 July 2006.

Hyun, I., Hochedlinger, K., Jaenisch, R., and Yamanaka, S. (2007) 'New Advances in iPS Cell Research Do Not Obviate the Need for Human Embryonic Stem Cells', *Cell Stem Cell*, 1, 11 October, 367–8.

Iliadou, E. (1999) *Forschungsfreiheit und Embryonenschutz. Eine verfassungs- und europarechtliche Untersuchung der Forschung an Embryonen* (Berlin: Duncker and Humblot).

Indian Council of Medical Research (ICMR) (2004) *Draft Guidelines for Stem Cell Research Regulation in India*, http://www.icmr.nic.in/bioethics/guidelines_stemcell.pdf.

Independent (1999) 25 June.

International Society of Bioethics, Scientific Committee (ISB) (2000) *Bioethics Declaration of Gijon*, http://www.sibi.org/ingles/declaracion.htm.

Israel Academy of Sciences and Humanities Bioethics Advisory Committee (2001) *The Use of Embryonic Stem Cells for Therapeutic Research*, http://www.academy.ac.il/bioethics/english/main-e.html.

Jacobzone, S. (2003) 'Healthy Ageing and the Challenges of New Technologies: Can OECD Social and Health-Care Systems Provide for the Future?', in OECD (ed.), *Biotechnology and Healthy Ageing: Policy Implications of New Research* (Paris: OECD), 37–54.

Jasanoff, S. (2005a) 'Democracies of DNA: The Politics of Bioethics', Paper Presented to the Bioethics Seminar, University of Sydney, Sydney, 17 August, Australia.

Jasanoff, S. (2005b) *Designs on Nature* (Princeton, NJ: Princeton University Press).

Jayaraman, K. S. (2005) 'Indian Regulations Fail to Monitor Growing Stem-Cell Use in Clinics', *Nature*, 434, 259.

Jessop, B. (2002) *The Future of the Capitalist State* (Oxford: Polity Press).

Jonsen, A. (1998) *The Birth of Bioethics* (New York: Oxford University Press).

Jonsen, A. R. (1994) 'Foreword', in E. R. Dubose, R. P. Hamel, and L. J. O'Connell, *A Matter of Principles? Ferment in US Bioethics* (Valley Forge, PA: Trinity Press International).

Jütte, R. (1993) *Geschichte der Abtreibung. Von der Antike bis zur Gegenwart* (München: C. H. Beck).

Kahn, J. (2000) 'Making a Market for Human Embryos', CNN.com, posted 4 September, date accessed 23 January 2006.

Kamal, A. and Hinsliff, G. (2000) 'Human Embryos to Be Cloned: Church Fury over "Spare Parts" Research', *The Observer*, 30 July.

Kapadia, C. R. (2000) Comments on stem cell guidelines, by Cyrus R. Kapadia, School of Medicine, Yale University, January 24 (Letter to the NIH).

Kazuto, K. (2005) 'The Ethical and Political Discussions on Stem Cell Research in Japan', in W. H. Bender and Christine Alexandrea Manzel (eds), *Grenzüberschreitungen: Kulturelle, religiöse und politiche Differenzen im Kontext der Stammzellenforschung weltweit* (Munich: Agenda Verlag), 369–79.

Keates, T. (1995) '*Bioethics in Italy*', *The Lancet*, 345(21), January, 182.

Keefe, B. (2004) 'Stem Cell Bond Issue Could Boost California Biotech Role', *Cox News Service*, 24 October, http://www.dailysentinel.com/health/content/shared/news/stories/CALIF_STEMCELLS__ADV24_COX.html, date accessed 25 July 2005.

Keller, E. F. (2000) *The Century of the Gene* (Cambridge, MA: Harvard University Press).

Kelly, S. (2003) 'Public Bioethics and Publics: Consensus, Boundaries and Participation in Biomedical Science Policy', *Science, Technology and Human Values*, 28(3), 339–64.

Kennedy, D. (2006) 'Editorial retraction', *Sciencemag*, 20 January, http://www.sciencemag.org/cgi/reprint/311/5759/335b.pdf, date accessed 18 July 2007.

Khushf, G. (1997) 'Embryo Research: The Ethical Geography of the Debate', *Journal of Medicine and Philosophy*, 22, 495–519.

Kim, H. K. (2005) *Biotechnology and Politics* (Seoul: Whistler) (in Korean).

Kim, T. G. (2006) '$450 Million Budget Set for Stem Cell Research', *Korean Times*, 29 May.

Kimbrell, A. (1997) *The Human Body Shop: The Cloning, Engineering, and Marketing of Life*, 2nd edition (Washington, DC: Regnery).

Kinner, J. K. (2000a) 'Regulating Embryonic Stem Cell Research: Biomedical Investigation of Human Embryos', in National Bioethics Advisory Commission (ed.), *Ethical Issues in Human Stem Cell Research* (vol. 2) (Rockville, MD), G1–G19.

Kinner, J. K. (2000b) 'Bioethical Regulation of Human Fetal Tissue and Embryonic Germ Cellular Material: Legal Survey and Analysis', in National Bioethics Advisory Commission (ed.), *Ethical Issues in Human Stem Cell Research* (Rockville, MD), F1–F27.

Knoppers, B. M. and Chadwick, R. (1994) 'The Human Genome Project: Under an International Ethical Microscope', *Science*, 265, 2035–6.

Knoppers, B. M. and Chadwick, R. (2005) 'Human Genetics Research: Emerging Trends in Ethics', *Nature Reviews Genetics*, 6, 75–9.

Koch, T. (2006) 'Bioethics as Ideology: Conditional and Unconditional Values', *Journal of Medicine and Philosophy*, 31, 251–67.

Korean National Commission for UNESCO (KNCU) (1999) *The 2nd Korean Consensus Conference on Cloning*, http://www.unesco.or.kr/cc/eng.html.

Kumar, T. C. (2004) '*In Vitro* Fertilization in India', *Current Science*, 86 (2, 25), 254–6.

Kumaranayake, L. and Lake, S. (2002) 'Regulation in the Context of Global Health Markets', in K. Lee, K. Buse, and S. Fustukian (eds), *Health Policy in a Globalising World* (Cambridge: Cambridge University Press), 78–96.

Laclau, E. (1990) *New Reflections on the Revolution of Our Time* (London: Verso).

Laclau, E. and Mouffe, C. (1985) *Hegemony and Socialist Strategy: Toward a Radical Democratic Politics* (London: Verso).

LaFleur, W. R. (1992) *Liquid Life: Abortion and Buddhism in Japan* (Princeton, NJ: Princeton University Press).

Lanza, R. (2007) 'Stem Cell Breakthroughs: Don't Forget the Ethics', *Science*, 318 (December 21), 1865.

Lanzendorf, S. E., Boyd, C. A., Wright, D. L., Muasher, S. Oehninger, S. and Hodgen, G. D. (2001) 'Use of Human Gametes Obtained from Anonymous Donors for the Production of Human Embryonic Stem Cell Lines', *Fertility and Society*, 76, 132–7.

Latour, B. (1987) *Science in Action: How to Follow Scientists and Engineers through Society* (Cambridge, MA: Harvard University Press).

Latour, B. (1988) *The Pasteurization of France*, trans. by A. Sheridan and J. Law (Cambridge, MA: Harvard University Press).

Lee, K. (1999) 'The Fear of Cloning', *Current Biology* 9 (August), 56.

Leeper, E. M. (1975) 'Fetal Research: Commission Sets Guidelines for Experimentation', *Bioscience* 25 (June), 357–60.

Leshner, A. I. and Thomson, J. A. (2007) 'Standing in the Way of Stem Cell Research', *Washington Post*, 3 December, A17.

Lock, M. (2001) 'The Alienation of Body Tissue and the Biopolitics of Immortalized Cell Lines', *Body and Society*, 7(2–3), 63–91.

Löfgren, H. and Benner, M. (2005) 'The Political Economy of the New Biology: Biotechnology and the Competition State', Paper Presented at the Druid Tenth Anniversary Summer Conference on Dynamics of Industry and Innovation: Organizations, Networks and Systems, Copenhagen Business School, 27–9 June, Copenhagen.

Lopez, J. (2004) 'How Sociology Can Save Bioethics ... Maybe', *Sociology of Health and Illness*, 25(7), 875–96.

Los Angeles Times (2001) 'Stem Cell Study Decision by Summer', March 1, p. 12.

Magnus, D. and Cho, M. (2005) 'Issues in Oocyte Do nation for Stem Cell Research', *Science*, 308, 1747–8.

Magureanu, G. (2005) Presentation to CORE *European Seminar: Human Egg Trading and the Exploitation of Women*. European Parliament, 30 June.

Maingueneau, Dominique (1984) *Genèses du Discourse* (Liége: Mardaga).

Majone G. (1989) *Evidence, Argument and Persuasion in the Policy Process* (New Haven, CT: Yale University Press).

Mark, M. (2002) 'Davis OKs Stem Cell Research. California Is First State to Encourage Studies', *San Francisco Chronicle*, Chronicle Sacramento Bureau, 23 September.

Martins, J., Gonand, F., Antolin, P., de la Maisonneuve, C., and Yoo Kwang-Yeol (2005) *The Impact of Ageing on Demand, Factor Markets and Growth*. Economics Working Papers No. 420 (Paris: OECD).

Masood, E. (1999) 'Expert Group to Look at UK Cloning Law', *Nature*, 400, 4.

Mayumi, M. (2006) 'Present State of Reproductive Medicine in Japan – Ethical Issues with a Focus on Those Seen in Court Cases', *BMC Medical Ethics*, 7(3), 1–16.

McConkey, D. (2001) 'Whither Hunter's Culture War? Shifts in Evangelical Morality, 1988–1998', *Sociology of Religion*, 62, 149–74.

McCook, A. (2007) 'Stem Cell Patents Loosened', *The Scientist Online*, 23 January, www.the-scientist.com/news/display/43099/, accessed 12 November 2007.

McKay, R. (2000) 'Stem Cells – Hype and Hope', *Nature*, 406, 361–4.

McLaren, A. (2000) 'Cloning: Pathways to a Pluripotent Future', *Science*, 288 (5472), 1775–80.

McMeekin, A. and Green, K. (2002) 'The Social and Economic Dimensions of Biotechnology: An Introduction', *New Genetics and Society*, 21(2), 101–8.

Mowery, D. and Sampat, B. (2004) 'The Bayh-Dole Act of 1980 and University–Industry Technology Transfer: A Model for Other OECD Governments?', *Journal of Technology Transfer*, 30(1–2), 115–27.

Ministero della Sanità (1997a) '*Ordinanza, 5 marzo 1997, Divieto di commercializzazione e di pubblicità di gameti ed embrioni umani*', Gazzetta Ufficiale 7 March, n. 55.

Ministero della Sanità (1997b) '*Ordinanza, 5 marzo 1997, Divieto di pratiche di clonazione umana o animale*', Gazzetta Ufficiale 7 March, n. 55.

Ministero della Sanità, U. S. (2000) 'TNSA: la via italiana all'uso delle cellule staminali' Comunicato [stampa] numero 37, 28 December 2000.

Ministero della Sanità (2003 [2000]) 'Relazione della Commissione di studio sull'utilizzo di cellule staminali per finalità terapeutiche', Roma.

Minkenberg, M. (2002) 'Religion and Public Policy: Institutional, Cultural and Political Impact on the Shaping of Abortion Policies in Western Democracies', *Comparative Political Studies*, 35(2), 221–47.

Mitchell, S. (2002), *United Press International (UPI)*, 7 November.

Mulkay, M. (1997) *The Embryo Research Debates* (Oxford: Oxford University Press).

Murray, F. (2007) 'The Stem-Cell Market — Patents and the Pursuit of Scientific Progress', *New England Journal of Medicine*, 356, 2341–3.

Nador, A. and Loucaides, T. (2003) *Stem Cells: Patents and Related Issues* (London: Berenskin and Parr).

Nahman, M. (2005) 'Israeli Extraction: An Ethnographic Study of Egg Donation and National Imaginaries' (Unpublished PhD Thesis, Lancaster University, UK).

Nakatsuji, N. (2007) 'Irrational Japanese Regulations Hinder Human Embryonic Stem Cell Research', *Nature Reports Stem Cells*, http://www.nature.com/stemcells/2007/0708/070809/full/stemcells.2007.66.html.

National Bioethics Advisory Commission, ed. (2000) *Ethical Issues in Human Stem Cell Research* (vols 1, 2) (Rockville, MD: Author).

National Bioethics Advisory Committee (NBAC) (1997) *Cloning Human Beings. Report and Recommendations of the National Bioethics Advisory Committee*, Rockville, MD, June.

National Bioethics Committee (NBC) (2006) *Interim Report on Ethical Issues of Professor Hwang's Research.*

National Commission for the Protection of Human Subjects of Biomedical and Behavioral Research (1978) The Belmont Report. Ethical Principles and Guidelines for the Protection of Human Subjects of Research (Washington, DC: GPO).

Nationaler Ethikrat (2001) 'Stellungnahme zum Import humaner embryonaler Stammzellen', 20 December.

National Institutes of Health (1999a, updated 2001) *NIH Fact Sheet on Human Pluripotent Stem Cell Research Guidelines*, Bethesda, MD.

National Institutes of Health (1999b) Draft National Institutes of Health Guidelines for Research Involving Human Pluripotent Stem Cells (Federal Register 67576), Bethesda, MD.

National Institutes of Health (2000a) *Stem Cells: A Primer*, Bethesda, MD.

National Institutes of Health (2000b) *Guidelines for Research Using Human Pluripotent Stem Cells* (Federal Register 51976), Bethesda, MD.

National Institutes of Health (2001a) *Notice of Criteria for Federal Funding of Research on Existing Human Embryonic Stem Cells and Establishment of NIH Human Embryonic Stem Cell Registry* (NOT-OD-02-005), Bethesda, MD.

National Institutes of Health (2001b) *Update on Existing Human Embryonic Stem Cells*, 27 August, Washington, DC.

Nature Editorial (2006) 'Safeguards for Donors', *Nature*, 442 (7103), 601.

Nature Editorial (2008) 'New Sources of Sex Cells', *Nature*, 452 (7190), 913.

Nature News (2005) 'Licensing Fees Slows Advance of Stem Cells', *Nature*, 435, 272–3.

Neilson, B. (2003) 'Globalization and the Biopolitics of Aging', *CR: The New Centennial Review*, 3(2), 161–86.

Nelkin, D. and Lindee, M. S. (1995) *The DNA Mystique: The Gene as a Cultural Icon* (New York: Freeman).

Neresini, F. (2000) 'And Man Descended from the Sheep: The Public Debate on Cloning in the Italian Press', *Public Understanding of Science*, 9, 359–82.

Novas, C. and Rose, N. (2000) 'Genetic Risk and the Birth of the Somatic Individual', *Economy and Society*, 29(4), 485–513.

Nuffield Council on Bioethics (2000) *Stem Cell Therapy: The Ethical Issues*. A Discussion Paper, London.

Nuffield Council on Bioethics (2002) *The Ethics of Patenting DNA* (London: Nuffield Council on Bioethics).

O'Connor, S. M. (2005) 'Intellectual Property Rights and Stem Cell Research: Who Owns the Medical Breakthroughs?', *New England Law Review*, 35, 665–714.

OECD (1989) *Biotechnology and Wider Impacts* (Paris: OECD).

OECD (2003) *Biotechnology and Healthy Ageing: Policy Implications of New Research* (Paris: OECD).

OECD (2004) *Biotechnology for Sustainable Growth and Development* (Paris: OECD).

OECD (2006) *The Bioeconomy to 2030: Designing a Policy Agenda* (Paris: OECD).

Ohmae, K. (1995) *The End of the Nation State* (New York: Free Press).

Padma TV (2005) 'India Plans Stem Cell Initiative', *SciDevNet*, 13 January, http://www.scidev.net/News/index.cfm?fuseaction=readnews&itemid=1849&language=1, date accessed 9 May 2007.

Paik, Young-Gyung (2006) 'Beyond Bioethics: The Globalized Reality of Ova Trafficking and the Possibility of Feminist Intervention', Paper Presented at *The International Forum on the Human Rights of Women and Biotechnology*, 21 September Seoul Women's Plaza, Seoul, Korea.

Pasotti, J. and Stafford, N. (2006) 'It's Legal: Italian Researchers Defend Their Work with Embryonic Stem Cells. Scientists Respond to Cardinal's Call for Excommunication', *Nature*, 442 (July 20), 229.

Peel Report, The (1972) Chairman Sir John Peel. *The Use of Fetuses and Fetal Material for Research* (London: HMSO).

People Science and Policy (2003) *Public Consultation on the Stem Cell Bank: Report Prepared for the Medical Research Council* (London: People Science & Policy Ltd).

Perrin, N. (2005) 'The Global Commercialisation of UK Stem Cell Research' (London: Report for UK Trade & Investment).

Polke-Majewski, K. (2002) 'Mutiger Anfang für die Biopolitik', *FAZ, Frankfurter Algemeine Zeitung*, 31 January.

Polkinghorne Report, The (1989) Chairman Rev. Dr John Polkinhorne. *Review of the Guidance on the Research Use of Fetuses and Fetal Material* (London: HMSO).

Pollock, A. (2003) 'Complicating Power in High-Tech Reproduction: Narratives of Anonymous Paid Egg Donors', *Journal of Medical Humanities*, 24(3/4), 241–63.

Pontifical Academy for Life (2000) *Declaration on the Production and Scientific and Therapeutic Use of the Human Embryonic Stem Cells*, 25 August (Rome: The Vatican), http://www.cin.org/docs/stem-cell-research.html, date accessed 9 September 2005.

Porter, G., Denning, C., Plomer, A., Sinden, J., and Torremans, P. (2006) 'The Patentability of Human Embryonic Stem Cells in Europe', *Nature Biotechnology*, 24, 653–5.

Prainsack, B. (2004) 'Negotiating Life: The Biopolitics of Embryonic Stem Cell Research and Human Cloning in Israel', Paper Presented at 4S/EASST Conference, Ecole de Mines, 24–9 August, Paris.

Prainsack, B. (2005) '"Negotiating Life": The Regulation of Human Cloning and Embryonic Stem Cell Research in Israel', *Social Studies of Science*, 34, 1–33.

Presidential Documents (US) (1995), Executive Order 12975 of 3 October 1995, Protection Of Human Research Subjects and Creation of National Bioethics Advisory Commission Federal Register: 5 October 1995 (vol. 60, no. 193), pp. 52063–5.

Presidential Documents (US) (1997), Weekly Compilation of Presidential Documents, pp. 845–6, Monday, 16 June 1997 (vol. 33, no. 24), pp. 843–70. Message to the Congress Transmitting the Proposed 'Cloning Prohibition Act of 1997', 9 June 1997.

President's Council on Bioethics (PCB) (US) (2002) *Human Cloning and Human Dignity: An Ethical Inquiry*, Washington, DC, 11 July.

President's Council on Bioethics (PCB) (2004) *Monitoring Stem Cell Research* (Washington, DC: President's Council on Bioethics).

President's Council on Bioethics (PCB) (2005) *Alternative Sources of Human Pluripotent Stem Cells*. White Paper (Washington, DC: President's Council on Bioethics).

ProLife Alliance (2001) Manifesto 2001, Belfast.

Rabb, H. (1999) Letter from HHS Gen. Counsel Harriet Rabb to Harold Varmus, Director, NIH, 15 January.

Rabin, S. (2005) 'The Gatekeepers of hES Cell Products', *Nature Biotechnology*, 23, 817–19.

Rabinow, P. and Rose, N. (2006) 'Biopower Today', *BioSocieties*, 1, 195–217.

Radford, T. (2000) 'MPs To Get Free Vote on Embryo Cures', *The Guardian*, 17 August, www.guardian.co.uk, date accessed 31 August 2007.

Rau, J. (2001a) *Rede bei der Sondersitzung des Deutschen Bundestages aus Anlass des Gedenktages fuer die Opfer des Nationalsozialismus* (Speech given in Berlin, Germany, 26 January).

Rau, J. (2001b)*Wird alles gut? Für einen Fortschritt nach menschlichem Maß* (Speech given in Berlin, Germany, 18 May).

Rau, J. (2001c) *Wird alles gut? Für einen Fortschritt nach menschlichem Maß* (Frankfurt: Suhrkamp Verlag).

Red Herring (2004) 'The Stem Cell Research Gold Rush', 1 November, http://www.redherring.com/Article.aspx?a=10941&hed=The+stem+cell+research+gold+rush, date accessed 27 July 2005.

Reich, W. T. (1995) 'The Word "Bioethics": The Struggle over Its Earliest Meanings', *Kennedy of Ethics Journal*, 5, 19–34.

Repubblica Italiana (2004) 'Legge 19 febbraio 2004, n. 40, Norme in materia di procreazione medicalmente assistita', *Gazzetta Ufficiale della Repubblica Italiana, Serie generale*, 24 February.

Research Europe (2003) 'Eu Stem Cell Guidelines Raise Supply Issues'. *Research Europe*, 31 July, 4.

Research Fortnight (2006) 'Stem Cell Compromise Allows Approval of FP7 by Council', 25 July, http://cordis.europa.eu/search/index.cfm?fuseaction=news.simpledocument&N_RCN=26062, date accessed 10 July 2007.

Reynolds, J. and Darnovsky, M. (2006) *The California Stem Cell Program at One Year: A Progress Report. Center for Genetics and Society.* January, www.genetics-and-society.org. accessed 14 July 2007.

Robertson, J. A. (1999) 'Ethics and Policy in Embryonic Stem Cell Research', *Kennedy Institute of Ethics Journal*, 9(2), 109–36.

Rose, Nikolas (2001) 'The Politics of Life Itself', *Theory, Culture & Society*, 18(6), 1–30.

Rose, N. and Novas, C. (2005) 'Biological Citizenship', in A. Ong and S. Collier (eds), *Global Assemblages: Technology, Politics, and Ethics as Anthropological Problems* (Oxford: Blackwell Publishing), 439–63.

Rosenberg, C. E. (1999) 'Meanings, Policies and Medicine: On the Bioethical Enterprise and History', *Daedelus*, 128(4), 27–46.

Rosenbladt, S. (1988) *Biotopia: Die Genetische Revolution und ihre Folgen für Mensch, Tier und Umwelt* (Berlin: Knaur Verlag).

Rothman, D. (1991) *Strangers at the Bedside: A History of How Law and Bioethics Transformed Medical Decision Making* (London: Basic Books).

Rothstein, H., Irwin, A., Yearley, S., and McCarthy, E. (1999) 'Regulatory Science, Europeanization and the Control of Agrochemicals', *Science, Technology and Human Values*, 24(2), 241–64.

The Royal Society (2000a) *Stem Cell Research and Therapeutic Cloning: An Update*, London, November.

The Royal Society (2000b) *Therapeutic Cloning: A Submission by the Royal Society to the Chief Medical Officer´s Expert Group*, February.

Sack, K. and Niebuhr, G. (2001) 'After Stem-Cell Rift, Groups Unite for Anti-abortion Push', *New York Times*, 4 September.

Salter, B. (2005) 'The Global Politics of Human Embryonic Stem Cell Science', Paper Presented at the ECPR Conference, 14–19 April, Grenada, Spain.

Salter, B. (2006a) 'Bioethics, Patenting and the Governance of Human Embryonic Stem Cell Science: the European Case', Working Paper 8, Global Biopolitics Research Group, University of East Anglia.

Salter, B. (2006b) 'Cultural Politics and Human Embryonic Stem Cell Science', in A. Webster, and S. Wyatt (eds), *New Technologies in Health Care* (Basingstoke: Palgrave Macmillan), 211–23.

Salter, B. and Jones, M. (2002) 'Human Genetic Technologies, European Governance and the Politics of Bioethics', *Nature Reviews Genetics*, 3, 808–14.

Salter, B. and Jones, M. (2005) 'Biobanks and Bioethics: The Politics of Legitimation', *Journal of European Public Policy*, 12(4), 710–32.

Saltman, R. (2003) 'Melting Public-Private Boundaries in European Health Systems', *The European Journal of Public Health*, vol. 13, no. 1.

Sanides, S. (2003) 'Stem Cell Pioneer Oliver Bruestle's Work Prompts New Law for German Scientists', *The Scientist*, 17, 52.

Sassen, S. (2002) 'Global Cities and Survival Circuits', in B. Ehrenreich and R. Hochschild (eds), *Global Woman: Nannies, Maids and Sex Workers in the New Economy* (London: Granta Books), 254–74.

Satolli, R. and Terragni, F. (eds) (1998) *La clonazione e il suo doppio* (Milano: Garzanti).

Schaffer, Simon and Shapin, Steven (1989) *Leviathan and the Air-Pump* (Princeton: Princeton University Press).

Scheil-Adlung, X. (1998) Zur Politik der Steuerung von Gesundheitsausgaben durch Verhaltensanreize: Eine vergleichende Analyse ausgewählter OECD Ländern. *Internationale Revue für Soziale Sicherheit*, Bd. 51 (1), 1998, 115–52.

Schepel, H. (2005) *The Constitution of Private Governance: Product Standards in the Regulation of Integrated Markets* (Oxford and Portland: Hart Publishing).

Scheper-Hughes, N. (2000) 'The Global Traffic in Human Organs', *Current Anthropology*, 41(2), 191–224.

Scheper-Hughes, N. (2002) 'Bodies for Sale: Whole or in Parts', in N. Sheper-Hughes and L. Wacquant (eds), *Commodifying Bodies* (London: Sage Publications), 1–8.

Schlieter, J. (2005) 'Stammzellenforschung und -therapie aus Sicht der buddhistischen Ethik', in W. H. Bender, Christine and Alexandrea Manzel (eds),

Grenzüberschreitungen: Kulturelle, religiöse und politiche Differenzen im Kontext der Stammzellenforschung weltweit (Münster: Agenda Verlag), 185–201.

Schlieter, J. (2005) 'Stammzellenforschung und therapie aus Sicht der buddihistischen Ethik', in W. H. Bender and Christine Alexandrea Manzel (ed.), *Grenzüberschreitungen: Kulturelle, religiöse und politiche Differenzen im Kontext der Stammzellenforschung weltweit* (Münich: Agenda Verlag), 185–201.

Schmiese, W. (2000) ' ... und Deutschland folgt?', *Die Welt*, December 21.

Schneider, I. (1995) *Föten als neuer medizinischer Rohstoff* (Frankfurt am Main: Campus Verlag).

Schröder, Gerard (2001) Rede des Bundekanzler zur Konstituierenden Sitzung des Nationale Ethikrates, Bulletin der Bundesregierung, 39–3, 8 June, Presse und Informationsamt der Bundesregierung, Bonn.

Science and Technology Committee (1996–7), Fifth Report of Session (HC373-I).

The Scientist (2004) 'Big Changes Afoot in Spain', 29 March, http://www.biomedcentral.com/news/20040329/02/.

Sell, S. K. (2005) 'The Doha Development Agenda: Intellectual Property', *Endgame at the World Trade Organisation: Reflections on the Doha Development Agenda*, Workshop, 11–12 November, European Research Institute, Birmingham.

Sexton, S. (2005) 'Transforming "Waste" into "Resource": From Women's Eggs to Economics for Women', Reprokult workshop, *Femme Globale* Conference, Heinrich Böll Foundation, 10 September, Berlin, Germany.

Singapore Bioethics Advisory Committee (2002) *Ethical, Legal and Social Issues in Human Stem Cell Research, Reproductive and Therapeutic Cloning*, www.bioethicssingapore.org/resources/body_reports1.html.

Snyder, E. and Vescovi, A. (2000) 'The Possibilities/Perplexities of Stem Cells', *Nature Biotechnology*, 18, 827–8.

Spallone, P. (1999) 'How the Pre-Embryo Got Its Spots', Einstein Forum's International Conference on Genetics and Genealogy, 4–6 July, Potsdam.

Sperling, S. (2004) 'From Crisis to Potentiality; Managing Potential Selves: Stem Cells, Immigrants, and German Identity', *Science and Public Policy*, 31(2), 139–49.

Starnge, S. (2000) *The Retreat of the States: The Diffusion of Power in the World Economy* (Cambridge: Cambridge University Press).

Steering Committee of the International Stem Cell Initiative (2005) 'The International Stem Cell Initiative: Toward Benchmarks for Human Embryonic Stem Cell Research' *Nature Biotechnology*, 23, 795–7.

Steinbrook, R. (2006) 'Egg Donation and Human Embryonic Stem-Cell Research', *New England Journal of Medicine*, 354(4), 324–6.

Stem Cell Business News (2007) 'Top Stem Cell Scientists Argue against Thomson/ WARF Patents', *Stem Cell Business News*, 5 July.

St John, J. and Lovell-Badge, R. (2007) Human-Animal Cytoplasmic Hybrid Embryos, Mitochondria, and an Energetic Debate', *Nature Cell Biology*, 9, 988–92.

Street, B. (1993) 'Culture Is a Verb: Anthropological Aspects of Language and Cultural Process' in D. Graddol, L. Thompson, and M. Byram (eds), *Language and Culture* (Clevedon, Avon: British Association for Applied Linguistics in association with Multilingual Matters), 23–43.

Tae-gyu, K. (2005) 'Stem Cell and Buddhism', *Korean Times*, 24 May, www.koreatimes.co.kr, date accessed 31 August 2007.

Takahashi, Kazutoshi, Tanabe, Koji, Ohnuki, Mari, Narita, Megumi, Ichisaka, Tomoko, Tomoda, Kiichiro, and Yamanaka, Shinya (2007) 'Induction of Pluripotent Stem Cells from Adult Human Fibroblasts by Defined Factors', *Cell* (19 November).

Tauer, C. A. (1997) 'Embryo Research and Public Policy: A Philosopher's Appraisal', *Journal of Medicine and Philosophy* 22, 423–39.

Taylor, C., Bolton, E., Pocock, S., Sharples, L., Pedersen, R., and Bradley J. (2005) 'Banking on Human Embryonic Stem Cells: Estimating the Number of Donor Cell Lines Needed for HLA Matching' *The Lancet*, 366, 2019–25.

TAZ (Die Tageszeitung) (2000) 23 February, p. 7.

TAZ (Die Tageszeitung) (2001a) 5 May, p. 9.

TAZ (Die Tageszeitung) (2001b) 6 June.

TAZ (Die tageszeitung) (2002).

Thomson, James A. et al. (1998) 'Embryonic Stem Lines Derived from Human Blastocysts' *Science*, 282, 1145–7.

Timmermans, S. and Berg, M. (2003) *The Gold Standard: The Challenge of Evidence-based Medicine and Standardization in Health Care* (Philadelphia: Temple University Press).

Timmermans, S., Bowker, G., and Star, S. (1998) 'The Architecture of Difference: Visibility, Control and Comparability in Building a Nursing Interventions Classification' in M. Berg and A. Mol (eds), *Differences in Medicine: Unravelling Practices, Techniques, and Bodies* (Durham, NC: Duke University Press), 203–25.

Titmuss, R. (1997) *The Gift Relationship: From Human Blood to Social Policy*, A. Oakley and J. Ashton (eds) (London: LSE Books).

Torfing, J. (1999) *New Theories of Discourse: Laclau, Mouffe, and Žižek* (Oxford: Blackwell).

Tran, C. (2006) 'WARF Stem Cell Patents Challenged', *The Scientist Online*, 10 October, www.the-scientist.com, date accessed 12 November 2007.

UK Intellectual Property Office (2003) 'Inventions Involving Human Embryonic Stem Cells', http://www.ipo.gov.uk/patent/p-decisionmaking/p-law/p-law-notice/p-law-notice-stemcells.htm, date accessed 3 October 2007.

UK Patent Office (2003) 'UK Patent Office to Allow Certain Stem Cell Patents', UK Patent Office Notices, April, http://www.patent.gov.uk/patent/notices/practice/stemcells.htm, date accessed 13 January 2005.

UK Stem Cell Bank (2007) 'How Are IP Issues Dealt with between Depositors and Users?', http://www.ukstemcellbank.org.uk/faq12.html, date accessed 3 October 2007.

UK Stem Cell Initiative (2005a) *Report and Recommendations* (London: Department of Health).

UK Stem Cell Initiative (2005b). 'Global Positions in Stem Cell Research: China', http://www.advisorybodies.doh.gov.uk/uksci/global/china.htm, date accessed 9 May 2007.

UK Stem Cell Initiative (2005c) UK Stem Cell Initiative, http://www.advisorybodies.doh.gov.uk/uksci/index.htm (home page), date accessed August 2005.

United Nations (2002) *Abortion Policies: A Global Review* (United National Publications).

United Nations Educational and Cultural Organisation (UNESCO) (1997) *Universal Declaration on the Human Genome and Human Rights*. Adopted by the General Conference of UNESCO at Its 29th Session, 11 November, http://portal.unesco.

org/en/ev.php-URL_ID=13177&URL_DO=DO_TOPIC&URL_SECTION=201. html, date accessed 4 February 2008.

United Nations Educational and Cultural Organisation, International Bioethics Committee (UNESCO) (2001) *The Use of Embryonic Stem Cells in Therapeutic Research*, http://portal.unesco.org/shs/en/ev.php-url_id=2144&url_do=do_topic&url_section=201.html.

Urry, J. (2000) *Sociology Beyond Societies: Mobilities for the Twenty-first Century* (London and New York: Routledge).

US President's Council on Bioethics (2002) *Human Cloning and Human Dignity: An Ethical Inquiry*, www.bioethics.gov.

US President's Council on Bioethics (2004) *Monitoring Stem Cell Research*, www.bioethics.gov.

US President's Council on Bioethics (2005) *Alternative Sources of Human Pluripotent Stem Cells*, www.bioethics.gov.

US Senate (1997) 105th Congress, 1st Session, S. 368, bill to prohibit the use of federal funds for human cloning research, 27 February 1997, introduced by Mr Bond and Mr Ashcroft, http://www.un.org/esa/population/publications/abortion/, date accessed 8 September 2005.

US Senate (1998a) Human Cloning Prohibition Act (1998) (Placed on Calendar in Senate) S 1601 PCS, Calendar No. 30, 105th CONGRESS, 2nd Session, S. 1601, to amend title 18, United States Code, to prohibit the use of somatic cell nuclear transfer technology for purposes of human cloning. Sponsored by: Sen. Christopher Bond (R, Mo.), and co-sponsored by Sens. Bill Frist (R, Tenn.), MD; Trent Lott (R, Miss.); and Judd Gregg (R, N.H.), 3 February 1998.

US Senate (1998b) Prohibition on Cloning of Human Beings Act of 1998 (Placed on Calendar in Senate), S 1611 PCS, Calendar No. 305, 105th CONGRESS, 2nd Session, S. 1611, to amend the Public Health Service Act to prohibit any attempt to clone a human being using somatic cell nuclear transfer and to prohibit the use of Federal funds for such purposes, to provide for further review of the ethical and scientific issues associated with the use of somatic cell nuclear transfer in human beings, and for other purposes, 4 February 1998. Sponsored by: Sens. Dianne Feinstein (D, Calif.) and Edward Kennedy (D, Mass.).

US Senate (1998c) Human Cloning Prohibition Act – Motion to Proceed, Page S599, 11 February.

Van Epps, H. (2006) 'Singapore's Multibillion Dollar Gamble', *Journal of Experimental Medicine*, 203(5), 1139–42.

Veatch, R. (1991) *The Patient–Physician Relation* (Bloomington: Indiana University Press).

Verfaillie, C. M. (2002) 'Pluripotency of Mesenchymal Stem Cells Derived from Adult Marrow', *Nature* 418, 41–9.

Vogel, Gretchen (1999) 'Breakthrough of the Year: Capturing the Promise of Youth', *Science*, 286, 2238–9.

Vogel, G. (2002) 'Pioneering Stem Cell Bank Will Soon Be Open for Deposits', *Science*, 297(5588), 1784.

Vogel, G. (2004a) 'Human Cloning: Scientists Take Step toward Therapeutic Cloning', *Science*, 303(660), 937–9.

Vogel, G. (2004b) 'Stem Cell Claims Face Legal Hurdles', *Science*, 305 (5692), 1887.

Vogel, G. (2006) 'Ethical Oöcytes: Available for a Price', *Science*, 313, 155.

Wade, N. (2001) 'Scientists Divided on Limit of Federal Stem Cell Money', *New York Times*, 16 August.

Wade, N. (2002) 'New Stanford Institute Is to Study Stem Cells', *New York Times*, 12 December.

Waldby, C. (2002) 'Stem Cells, Tissue Cultures and the Production of Biovalue', *Health: An Interdisciplinary Journal for the Social Study of Health, Illness and Medicine*, 6(3), 305–23.

Waldby, C. (2006) 'Umbilical Cord Blood: From Social Gift to Venture Capital', *BioSocieties*, 1(1), 55–70.

Waldby, C. and Mitchell, R. (2006) *Tissue Economies: Blood, Organs and Cell Lines in Late Capitalism* (Durham, NC: Duke University Press).

Waldby, C., Rosengarten, M., Treloar, C., and Fraser, S. (2004) 'Blood and Bioidentity: Ideas about Self, Boundaries and Risk among Blood Donors and People Living with Hepatitis C', *Social Science and Medicine*, 59(7), 1461–71.

Walters, L. (2004) 'The United Nations and Human Cloning: A Debate on Hold', *Hastings Center Report*, January–February, 5–6.

Walters, R. (2004) 'Human Embryonic Stem Cell Research: An Intercultural Perspective', *Kennedy Institute of Ethics Journal*, 14(1), 3–38.

Warnock Report (1985) Mary Warnock. *A Question of Life. The Warnock Report on Human Fertilisation and Embryology* (Oxford: Basil Blackwell).

Washington Post (2001) 24 August.

Weksler, M. (2003) 'Rapporteur's Scientific Summary', *Biotechnology and Healthy Ageing: Policy Implications of New Research* (Paris: OECD).

Wellcome Trust (2002) 'Memo'.

Wintour, P. (2000) 'Whitehall Split on Cloning Decision', *The Guardian*, 31 July, www.guardian.co.uk, date accessed 31 August 2007.

White House Fact Sheet (2001) 9 August, http://www.whitehouse.gov/news/releases/2001/08/ 20010809-1.html.

Williams, C., Kitzinger, J., and Henderson, L. (2003) 'Envisaging the Embryo in Stem Cell Research: Rhetorical Strategies and Media Reporting of the Ethical Debates', *Sociology of Health & Illness*, 25(7), 793–**814**.

Wolpe, P. R. and McGee, G. (2001) '"Expert Bioethics" as Professional Discourse: The Case of Stem Cells', in S. Holland, K. Lebacqz, and L. Zoloth (eds), *The Human Embryonic Stem Cell Debate: Science Ethics and Public Policy* (Cambridge, MA: MIT Press), 185–96.

World Health Organization (WHO) (1997) *Proposed International Guidelines on Ethical Issues in Medical Genetics and Genetic Services*, http://www.who.int/ncd/hgn/hgnethic.htm.

World Health Organization (2003) 'Human Organ and Tissue Transplantation', Report by the Secretariat, 27 November, http://www.who.int/gb/ebwha.

World Health Organisation (WHO) (2005) Ethics and Health Department, http://www.who.int/ethics/en.

World Medical Association (WMA) (2002) *Declaration on Ethical Considerations Regarding Health Databases*, http://www.wma.net/e/policy/smacdatabasesoct2002.htm.

Wright, S. (1998) 'The Politicisation of Culture', *Anthropology Today*, 14(1), 7–15.

Yahoo! News (2004) 'Spain to Authorise Stem Cell Research', http://story.news.yahoo.com/news?tmpl=story&cid=.../spain_health_biotech_04092718073.

Yesley, M. (2005) 'What's in a Name? Bioethics – and Human Rights – at UNESCO', *Hastings Center Report*, 35, 8.

Yomiuri (2007a) http://www.yomiuri.co.jp/dy/national/20071226TDY02305. htm, accessed 15 January 2008.

Yomuiri(2007b)http://www.yomiuri.co.jp/dy/features/science/20071221TDY03104. htm, accessed 15 January 2008.

Yu, Junying et al. (2007) 'Induced Pluripotent Stem Cell Lines Derived from Human Somatic Cells', *Science*, 318(5858), 1917–20.

Yu, J., Vodyanik, M. A., Smuga-Otto, K., ntosiewicz-Bourget, J., Frane, J. L., Tian, S., Nie, J., Jonsdottir, G. A., Ruotti, V., Stewart, R., Slukvin, I. I., and Thomson, J. A. (2007) 'Induced Pluripotent Stem Cell Lines Derived from Human Somatic Cells', *Science* 318, 1917–20.

Zeller, C. (2005) 'Innovation Systems in Biotechnology in a Finance Dominated Accumulation Regime', Paper Presented at Association of American Geographers Annual Meeting, April 8, Denver.

Zola, I. K. (1972) 'Medicine as an Institution of Social Control', *Sociological Review* 20, 4, 487–504.

Index